Innovations of Green Process Engineering for Sustainable Energy and Environment

Topical Conference at the 2010 AIChE Annual Meeting

Salt Lake City, Utah, USA
7-12 November 2010

ISBN: 978-1-61782-145-5

Printed from e-media with permission by:

Curran Associates, Inc.
57 Morehouse Lane
Red Hook, NY 12571

CURRAN ASSOCIATES INC. proceedings.com

Some format issues inherent in the e-media version may also appear in this print version.

Printed by Curran Associates, Inc. (2011)

AIChE
3 Park Avenue
New York, NY 10016-5991

Phone: (203) 702-7660
Fax: (203) 775-5177

www.aiche.org

Additional copies of this publication are available from:

Curran Associates, Inc.
57 Morehouse Lane
Red Hook, NY 12571 USA
Phone: 845-758-0400
Fax: 845-758-2634
Email: curran@proceedings.com
Web: www.proceedings.com

TABLE OF CONTENTS

FOULING MONITORING TECHNIQUES FOR WATER TREATMENT MEMBRANE PROCESSES

Kuo-Lun Tung, Shih-Chieh Lai, Nien-Jung Lin and Yu-Ling Li

Department of Chemical Engineering and R&D Center for Membrane Technology, Chung Yuan Christian University, Taoyuan 320, Taiwan

Abstract

An integrated, online fouling monitoring technique that provides dynamic and real-time information of membrane fouling during a water treatment membrane process was developed. This technique has preliminarily been demonstrated for monitoring particulate fouling in a water treatment microfiltration process using a single photo sensor. The results of the online dynamic analysis suggest that this technique provides in situ measurement of fouling layer thickness, dynamic analysis of the fouling layer structure, and monitoring of water quality during the membrane filtration process.

Introduction

Membrane fouling and flux decline are undesirable, albeit inevitable, problems in membrane filtration processes for water treatment. The development of low fouling membranes, the design of high efficiency modules, the selection of optimal operational strategies, and several improvements made in peripheral control, monitoring, and cleaning techniques have been adopted to deal with these formidable obstacles. Unfortunately, the slow development of these methods has limited the competitiveness and subsequent acceptance of membrane water treatment processes in the last three decades. An early warning technique for the determination of fouling problems in water treatment membrane processes is crucial to the improvement of membrane process operation, as well as to the development of a fouling prevention strategy. Severe flux decline and process failure are usually observed if the quality of the produced water fails a given set of standards. This is the most expensive method for fouling monitoring, but is still widely adopted due to its practicality. Recently, researchers have become more aware of the importance of in situ, online monitoring of fouling growth. Several in situ monitoring techniques have been developed in the laboratory to reveal the growth mechanism of fouling layers in membrane filtration processes (Chen *et al*, 2004a; Chen *et al*, 2004b). Nevertheless, few process-oriented, reliable, and predictable online monitoring techniques for onsite fouling analysis and control exist. In this study, an integrated, online monitoring technique that provides dynamic and real-time information of a fouling process has been developed. This monitoring technique has been preliminarily demonstrated for water treatment membrane processes using a photosensor array. The application to a water treatment membrane spearation unit with a photosensor array to provides real time, 3D profiling of fouling layer thickness changes. Using sensor array that are able to scan membranes surface on demand, an integrated, online, and dynamic monitoring technique would be feasible for direct application to commercial membrane modules for water treatment. The results of the online dynamic analysis suggest that this technique provides in situ measurement of fouling layer thickness, dynamic analysis of the fouling layer structure, and monitoring of water quality during the membrane filtration process. Moreover, it is notable that many available, robust, fouling monitoring devices have never moved beyond laboratory status. Significant challenges remain to improve available monitoring techniques to effectively bridge the gap between the laboratory setting and field application.

Fouling Monitoring and Dynamic Analysis

The proposed integrated fouling monitoring technique requires a device for detecting the time course of either the permeate volume or of the fouling layer growth thickness. With the measured real time data of the permeate accumulation or of the fouling layer thickness variation, a dynamic analysis procedure based on mass and force balances of particulate deposition on the membrane surface can be applied to analyze the fouling layer structure in a water treatment membrane process. The dynamic analysis was originally proposed by Lu and Hwang (1993), and was modified to analyze the formation and compression of the particle fouling layer during microfiltration by Tung *et al.* (2008). This online dynamic monitoring technique requires information about the pre-estimated surface porosity of the fouling layer and the real time variation of the fouling layer thickness. The surface porosity of a fouling layer can be obtained by a low head filtration system (Tung *et al.*, 2008). The primary difference between the analysis originally proposed by Lu and Hwang (1993) and that in this study is that the permeate flux versus time data used for analysis has been replaced with real time fouling layer thickness variation data.

Experimental

A constant pressure filtration was adopted in this work. The slurry was agitated in the slurry supply tank with a magnetic disk and was kept well mixed by pumping the slurry at a moderate rate through pressure filter. In each experiment, the filtration pressure was supplied by compressed air and was adjusted to a given value by a pressure regulator. The stainless steel filter chamber has a filtration area of $8.45 \times 10^{-4} m^2$. An LED photointerrupt sensor developed in our previous work (Tung *et al.*, 2001) was installed in front of the filter chamber to measure the growing thickness of the fouling layer. The increase of the weight of filtrate was detected by a load cell and recorded on a laptop. In this study, constant pressure dead-end microfiltration experiment of crosslinked *S. cerevisiae* with a density of 1,150 kg/m^3 was conducted to examine the structure of fouling layer formed by particles. The *S. cerevisiae* diameter ranged from 4 to 5 μm was purchased from Sigma Chemical Co., and suspended in 0.86% NaCl physiological saline to prepare slurry of a concentration of 0.2 w/w%. The *S. cerevisiae* particles were treated with 0.8(w/w)% glutaraldehyde to form deactivated particles. Track-etched polycarbonate membrane with pore size of 0.8 μm purchased from Millipore Co. was used as a filter medium in this study. The cake surface porosity can be obtained by means of a low head filtration system proposed by Haynes (Lu, 1968) for micron-size particles.

Results and Discussion

Formation and compression of a fouling layer

In order to examine the validity of the proposed online dynamic analysis technique, a dead-end microfiltration experiment using a crosslinked *S. cerevisiae* suspension was conducted under a constant operating pressure of 0.61 kPa. The time course of the fouling layer thickness variation was obtained using the aforementioned experimental conditions, and is illustrated in **Figure 1**. The accuracy of the fouling layer thickness measurement by the adopted photointerrupt sensor has been confirmed in our previous work (Tung *et al.*, 2001). The surface porosity was estimated to be 0.56 using the low head filtration test. With this experimental information, the intrinsic properties of the fouling layer can be dynamically and quantitatively analyzed by online dynamic analysis. The important internal properties of a fouling layer include porosity and resistance to filtration. The porosity of the fouling layer is one of the most important quantities for estimating fouling resistance. **Figure 2** illustrates the predicted porosity distributions during the course of dead-end microfiltration. From the plots in this figure, it can be observed that the fouling layer is compressed, and the local porosity in the fouling layer decreases gradually, from a surface porosity of 0.56, to a bottom porosity of 0.51. For constant pressure dead-end microfiltration, the values of porosity near the membrane and the surface of the fouling layer are nearly constant, and a sharp decrease is observed in most of the region of the fouling layer. The decreased fouling layer porosity is primarily due to the rearrangement of particles and the compression of the fouling layer, which are caused by frictional drag from fluid and by the compounded foulant mass of deposited particles. Fouling layers are compressible, that is, they become more compact as the extent of their compression increases. The solid compressive pressure is a response to the compression of a fouling layer according to basic filtration theory (Ruth, 1935). In traditional filtration theory, the derivation of the filtration drag equations for rigid particle slurries assumes that particles are in point contact mode, and that compression occurs instantaneously. Under this assumption, a force balance can be obtained between the liquid pressure over the entire cross-section and the solid compressive pressure on the total mass within the porous layer as $P_s + P_L = \Delta P$, where ΔP is the applied pressure and P_s is the solid compressive pressure. **Figures 3 and 4** illustrate the curves of ε versus P_s and α versus P_s, as obtained from dynamic analysis. This dynamic analysis procedure enables us to obtain these basis data by knowing the fouling layer surface porosity and real time variation of the fouling layer thickness. In summary, the adopted dynamic procedure is an effective tool for not only designing the membrane filtration system, but also for predicting and monitoring system performance during plant operation.

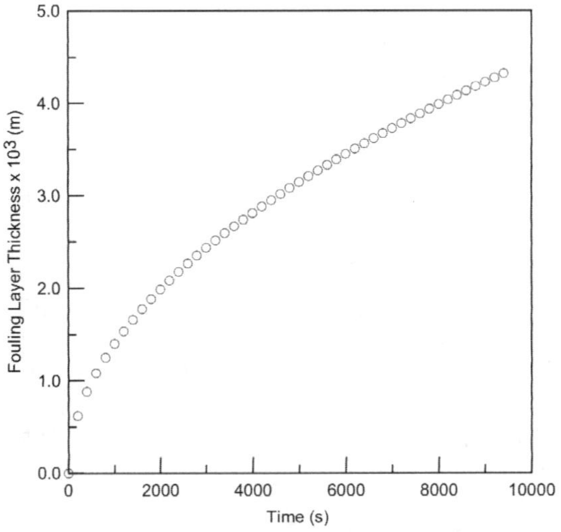

Figure 1 Time course of the fouling layer thickness measured at P=61 kPa for a 0.5%(w/w) crosslinked yeast suspension.

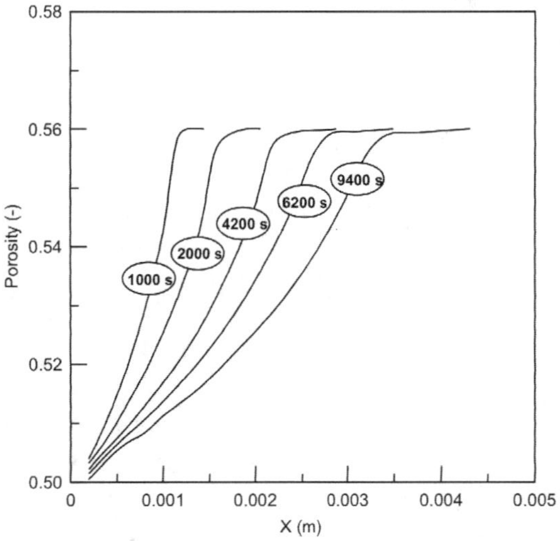

Figure 2 Analyzed porosity distributions in fouling layers during the time course of a dead-end microfiltration.

Monitoring water quality in a membrane filtration process

Figure 5 depicts the *t/V* versus *V* plot for the microfiltration of the crosslinked *S. cerevisiae* suspension. An almost linear filtration curve is observed in this plot, indicating that a cake filtration mode of filtration behavior occurs during the process. The time course of the MFI index during the microfiltration of the crosslinked *S. cerevisiae* suspension can be estimated from the slope of the filtration curve shown in **Figure 5** and is illustrated in **Figure 6**. Examination of the time course variation of the MFI index in **Figure 6** demonstrates that the operation was stable after 1 hour, as indicated by the almost constant MFI index value. The MFI index can also be used to monitor particulate fouling in water treatments and serve as predictor for the fouling potential of feed water containing particles (Alahmad, 2008).

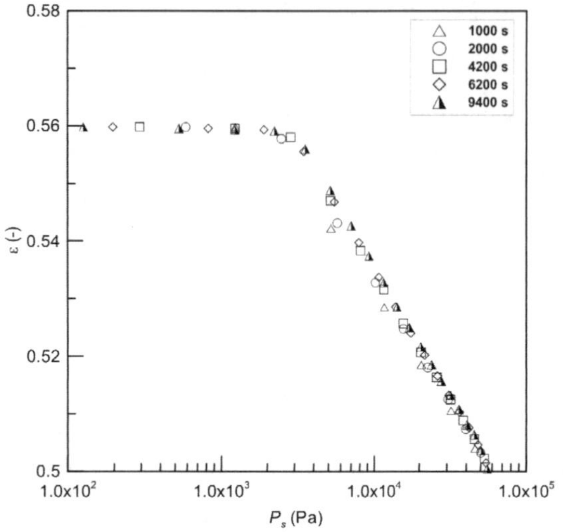

Figure 3 Porosity as a function of the solid compressive pressure (ε - P_s).

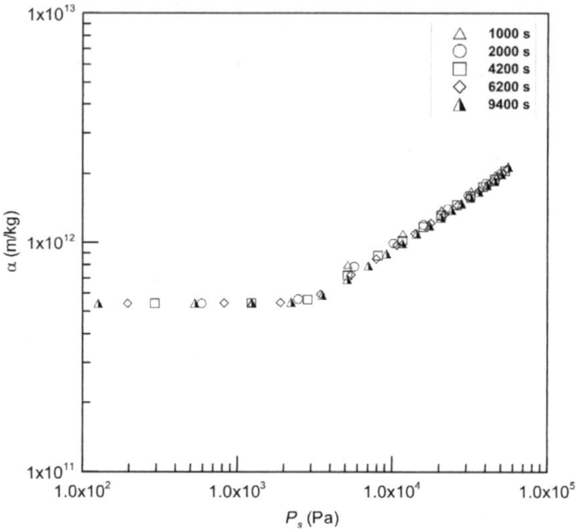

Figure 4 Specific filtration resistance as a function of the solid compressive pressure (α - P_s).

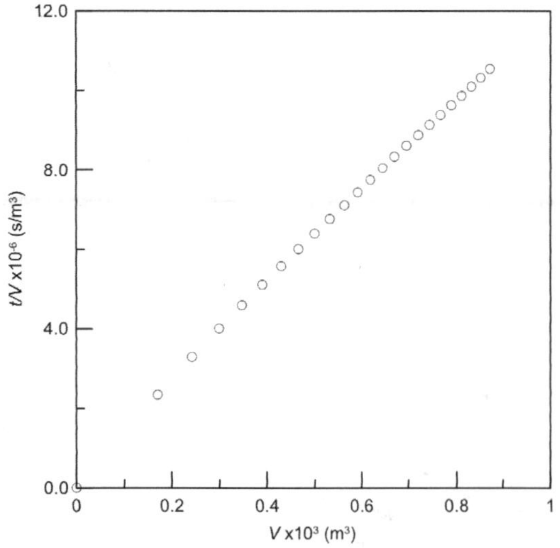

Figure 5 A plot of *t/V* versus *V* for the microfiltration of the

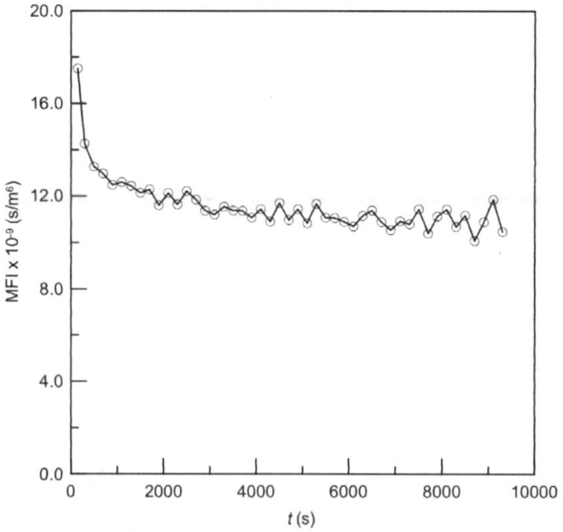

Figure 6 The time course of the MFI index estimated

crosslinked *S. cerevisiae* suspension.

during the MF of the crosslinked *S. cerevisiae* suspension by dynamic analysis.

A higher fouling potential feed water or suspension has a higher MFI value. A gradual increase in the MFI value indicates an increase in the filtration resistance, which may be caused by the compression of the fouling layer or the migration of fine particles into the fouling layer. An abrupt change in the MFI value during operation indicates an unexpected failure of the membrane filtration system, most likely caused by membrane or membrane module damage, or by faulty operation of the driving units. The MFI can be used in the laboratory as a quick test for feed quality and, in the long-term, to monitor the particulate fouling potential of a feed water or suspension.

Conclusions

In this study, an integrated, online monitoring technique that provides dynamic and real-time information of a fouling process has been developed. The results of the online dynamic analysis suggest that this technique provides in situ measurement of fouling layer thickness, dynamic analysis of the fouling layer structure, and monitoring of water quality during the membrane filtration process. The preliminary study represents one component of an integrated online dynamic monitoring technique, applied to a pilot-scale water treatment unit with a photo sensor array to provide real time profiling of fouling layer thickness changes. Using sensor array that is able to scan membranes surface on demand, an integrated, online, and dynamic monitoring technique would be feasible for direct application to commercial membrane modules for water treatment. Moreover, it is notable that many available, robust, fouling monitoring devices have never moved beyond laboratory status. Significant challenges remain to improve available monitoring techniques to effectively bridge the gap between the laboratory setting and field application.

References

1. Chen, J., Li, Q. and Elimelech, M. (2004a). *In situ* monitoring techniques for concentration polarization and fouling phenomena in membrane filtration. *Adv. Colloid Interface Sci.* **107**(2-3), 83-108.
2. Chen, V., Li, H. and Fane, A.G. (2004b). Non-invasive observation of synthetic membrane processes - a review of methods. *J. Membr. Sci.* **241**(1), 23-44.
3. Lu, W.M. (1968) *Theoretical and Experimental Analysis of Variable Pressure Filtration and the Effect of Side Wall Friction in Compression-Permeability Cells.* Ph.D. thesis, Univ. of Houston, Houston, Texas, USA.
4. Lu, W.M. and Hwang, K.J. (1993). Mechanism of cake formation in constant pressure filtrations. *Separa. Technol.* **3**(3), 122-132.
5. Tung, K.L., Wang, S., Lu, W.M. and Pan, C.H. (2001). *In situ* measurement of cake thickness distribution by a photointerrupt sensor. *J. Membr. Sci.* **190**(1), 57-67.
6. Tung, K.L., Li, Y.L., Hwang, K.J. and Lu, W.M. (2008). Analysis and prediction of fouling layer structure in microfiltration. *Desalination* **234**, 99-106.
7. B. F. Ruth, Studies in filtration, III derivation of general filtration equation *Ind. Eng. Chem* **1935,** 27, 708-723.

8. Alahmad, M., Factors affecting scale formation in sea water environments – an experimental approach. *Chem. Eng. Technol.* **2008**, 31(1), 149-156.

(50b) Characterization of Thin Film Composite (TFC) and Dense Membranes Based On Disulfonated Random Poly(arylene ether sulfone) Copolymers for Water Purification

Ozma Lane[1], Chang Hyun Lee[2], Wei Xie[3], Benny D. Freeman[4] and James McGrath[2], (1) Macromolecular Science and Engineering, Virginia Tech, Blacksburg, VA, (2)Chemistry, Virginia Tech, Blacksburg, VA, (3)Chemical Engineering, The University of Texas at Austin, Austin, TX, (4) Chemical Engineering, University of Texas at Austin, Austin, TX

Thin film composite and dense membranes were produced by casting a blend of BPS-20 (20% disulfonated poly(arylene ether sulfone)) or BPS-32 (32% disulfonated poly(arylene ether sulfone)) and varying molecular weights of poly(ethylene glycol) oligomer onto porous Udel polysulfone foam. An additional series of dense membrane and TFCs was produced from 35% disulfonated bisphenol A based copolymers. The thin skin thickness was evaluated using field emission scanning electron microscopy (FE-SEM) and the surface roughness was evaluated using tapping mode atomic force microscopy (AFM). The water flux and salt rejection values of the series were evaluated and compared with those of dense film membranes, as well as the skin thickness and roughness data. The addition of poly(ethylene glycol) in controlled compositions provided an increase in salt rejection.

See more of this Session: Research Frontier of Water Sustainability
See more of this Group/Topical: Topical G: Innovations of Green Process Engineering for Sustainable Energy and Environment

(50c) Membrane Filtration and Separation for Water Reuse in Fuel Production From Biomass

Wei Liu, Energy and Environmental Directorate, Pacific Northwest National Lab, Richland, WA

Water conservation is one of important sustainability issues in production and/or conversion of biomass to fuels. In this presentation, the author will use micro-algae growth and fermentation production of ethanol fuel as two examples to elaborate application of respective microfiltration and molecular separation membranes to water sustainability. Growth of micro-algae requires a large water resource, innovations of membrane materials and processes will be discussed for efficient harvesting of micro-algae from dilute culture solution and complete recycling of the water and nutrients. The fermentation process requires a lot of water injection and produces a dilute ethanol mixture. Studies on impacts of novel zeolite membranes on water reuses and minimization of waste water discharge will be presented.

See more of this Session: Research Frontier of Water Sustainability
See more of this Group/Topical: Topical G: Innovations of Green Process Engineering for Sustainable Energy and Environment

(50d) Water Treatment by Adsorption with Electrochemical Regeneration

Edward P. L. Roberts[1], Syed N. Hussain[1], Hafiz M. A. Asghar[1], Fadhil M Mohammed[1], Michael Conti-Ramsden[1], Nigel W. Brown[2] and Andrew K. Campen[2], (1)School of Chemical Engineering and Analytical Science, University of Manchester, Manchester, United Kingdom, (2)Arvia Technology Ltd, Chester, United Kingdom

An innovative process for water treatment using adsorption and electrochemical regeneration has been developed at the University of Manchester. A spin-out company, Arvia Technology Ltd. has been established to commercialise the technology. The process uses a graphite based adsorbent, Nyex® which has a low adsorptive capacity in comparison to many industrially used adsorbents. However, the material has a high electrical conductivity allowing electrochemical oxidation of adsorbed contaminants and 100% regeneration of its initial adsorptive capacity. It has been shown that a packed bed of Nyex® loaded with an organic contaminant can be regenerated in the anodic compartment of an electrochemical cell in around 10 minutes. Furthermore, the charge required for regeneration is approximately equal to the theoretical charge for complete oxidation of the organic. Unlike wastewater oxidation at a plate electrode, the breakdown products remain adsorbed on the surface of the Nyex® and are not released into solution. The rapid regeneration and low energy consumption make the process attractive for a wide range of water treatment applications. The process uses no chemicals and does not produce any solid or liquid wastes. Continuous and sequential batch treatment processes have been developed and these are currently being used in a series of industrial trials. In this paper, studies of the treatment of water contaminated with an organic dye, phenol, and mercaptans will be described, including analysis of any breakdown products present in the liquid and gas phases. In addition a preliminary study of the effectiveness of the process for disinfection will be reported.

See more of this Session: Research Frontier of Water Sustainability
See more of this Group/Topical: Topical G: Innovations of Green Process Engineering for Sustainable Energy and Environment

(50e) High Performance Dual-Layer Hollow Fiber Membranes for Forward Osmosis Processes

Jincai Su, Department of Chemical & Biomolecular Engineering, National University of Singapore, National University of Singapore, Singapore, Singapore and Tai-Shung Chung, Department of Chemical and Biomolecular Engineering, National University of Singapore, Singapore, Singapore

Forward osmosis (FO) is an energy effective technology and has great potentials in a variety of applications such as desalination of seawater, treatment of wastewater, concentration of protein or sugar, dehydration of alcohol, and generation of power. The understanding of this promising technology has been improved with intensive research works carried out worldwide. However, the lack of FO membranes with high water flux and high salt rejection has been a critical factor limiting the real application of this technology in industry.

Ideally, FO membranes should be semi-permeable and hydrophilic, allowing water to permeate through while rejecting other components. For this purpose, we have designed and successfully fabricated nanofiltration-based single-layer hollow fiber membrane from cellulose acetate (CA). However, severe concentration polarization occurs at the lumen side of the fibers because of the highly porous sublayer. The occurrence of concentration polarization has caused drastic reduction in the effective driving force in FO tests and resulted in a water flux much lower than the expected value. The attempt to develop dual-layer hollow fiber membranes is a continuation of the previous work. The objective is to suppress or reduce the negative impact of concentration polarization at the lumen side by varying the sublayer structure and/or chemical properties. The first batch of dual-layer hollow fiber membrane (CA-I) is based on CA, i.e., both of the outer and inner layers formed by CA. Being very thin and dense, the outer layer acts as the semi-permeable barrier for solute separation. The inner layer is highly porous, providing enough mechanical strength as membrane support while creating very little resistance for water permeation. No delamination or interface barrier is observed between the outer and inner layers. In the FO experiments with 0.5–2.0 M MgCl2 draw solutions versus the outer layer, the first batch of CA-based dual-layer hollow fiber membrane shows water fluxes and salt leakages in the range of 12.69–36.75 L m-2 h-1 and 0.16–0.21 g m-2 h-1, respectively. Another two batches of dual-layer hollow fiber membranes (CA-II and CA-III) are under study. The CA-II membrane is fabricated with different porosity or thickness of the support layer and the purpose is to optimizing the sublayer structure. In order to find out whether charge helps to reduce the concentration polarization at the lumen side, the CA-III membrane is fabricated with incorporating different percentages of sulfonated polysulfone into the sublayer. The FO experiments and characterization are being carried out on these two batches of dual-layer hollow fiber membranes and the results will be reported in due time.

See more of this Session: Research Frontier of Water Sustainability
See more of this Group/Topical: Topical G: Innovations of Green Process Engineering for Sustainable

(50f) Treatment of Produced Water Using Microbial Fuel Cells

Abhijeet P. Borole, Biosciences Division, Oak Ridge National Lab, Oak Ridge, TN and Costas Tsouris, Oak Ridge National Laboratory, Oak Ridge, TN

Sustainable utilization of natural energy sources in the future will require development of technologies with minimal impact on the environment. Produced water generated from oil and gas production operations currently generate 10 times more water than the amount of fossil fuel produced. In order to develop sustainable operation, the produced water needs to be either recycled or disposed into the environment as clean water. The current practice of produced water reinjection into the subsurface is not sustainable. Use of microbial fuel cells (MFCs) for enabling water recycle and reducing the energy required for treating produced water will be discussed. MFCs convert organic carbon and wastes into energy using naturally occurring and regenerable microbial catalysts. Effect of contaminants in produced water, namely salts and organics on MFCs and strategies to overcome them will be presented. As we look towards a future with restricted carbon release, while still being able to use fossil fuels, developing technologies minimizing impact on climate change and environment is of utmost importance. Reducing energy required for produced water treatment and enabling water reuse are two important components as we continue to use these resources in the future.

See more of this Session: Research Frontier of Water Sustainability
See more of this Group/Topical: Topical G: Innovations of Green Process Engineering for Sustainable Energy and Environment

(50g) Hydrophilic Superparamagnetic Nanoparticles: Synthesis, Characterization and Application in Forward Osmosis Processes

Qingchun Ge[1], Gary L. Amy[2] and Tai-Shung Chung[1], (1)Chemical and Biomolecular Engineering, National University of Singapore, Singapore, Singapore, (2)Water Desalination and Reuse Research Center, 4700 King Abdullah University of Science and Technology, Thuwal, Saudi Arabia
*Corresponding author, Email: chencts@nus.edu.sg, Fax: (65)-6779 1936

ABSTRACT

Freshwater shortage has caused serious world-wide problems and many technologies have been developed for water recycle and seawater desalination. Forward osmosis (FO) as a promising technology has attracted much attention, but the exploration of highly hydrophilic and easily separable draw solutes with high osmotic pressures makes this technology full of challenge.

Superparamagnetic nanoparticles are materials consisting of a magnetic core and polymer shell. The magnetic core contains small magnetic nanoparticles and can be separated from mixtures under an external magnetic field. The polymer shell surrounds the magnetic core to prevent particle aggregation and also enables modification. MNPs can be purposely designed and synthesized for specific applications. Such characteristics make magnetic nanoparticles an appropriate option of being draw solutes. In this work, we have synthesized a series of hydrophilic magnetic nanoparticles (MNPs), investigated their potential as draw solutes in FO systems, and explored their recovery and reusability. Good water fluxes are achieved in FO processes. The MNPs can be easily recovered from draw solutions by the magnetic field and reused. Results have demonstrated that using PEG based MNPs as draw solutes is a feasible way in the FO process.

See more of this Session: Research Frontier of Water Sustainability
See more of this Group/Topical: Topical G: Innovations of Green Process Engineering for Sustainable Energy and Environment

Cryogenic CO_2 Capture for Improved Efficiency at Reduced Cost

Christopher Hoeger[1], Chris Bence[1], Stephanie Stitt Burt[1], Andrew Baxter[1], Larry Baxter[1,2]
[1] Sustainable Energy Solutions, Provo, UT 84602
[2] Chemical Engineering Department, Brigham Young University, Provo, UT 84601

Abstract

Cryogenic carbon capture (CCC) is a carbon capture technology that has significant energy and cost advantages compared to alternatives. The process is a post-combustion technology that cools CO_2-laden flue gas to desublimation temperatures (-100 to -135 °C), separates solid CO_2 from the light gases in the flue gas, uses the cold products to cool incoming gases in a recuperative heat exchanger, compresses the solid/liquid CO_2 to final pressures (100-200 atm) read for sequestration or storage, and delivers a compressed CO_2 stream separated from an atmospheric pressure light-gas stream. The overall energy and economic costs appear to be at least 30% lower than most competing processes that involve air separation units (ASUs), solvents, or similar technologies. In addition, the CCC process enjoys several ancillary benefits, including (a) it is a minimally invasive bolt-on technology, (b) it provides highly efficient removal of most pollutants (Hg, SO_x, NO_2, HCl, etc.), (c) possible energy storage capacity, and (d) potential water savings. This paper outlines the process details and economic and energy comparisons relative to other well-documented alternatives.

Introduction

Energy and related environmental issues span national and regional boundaries and influence many generations. Their solutions require comparable scope and duration. Many people envision a future dominated by non-fossil energy generation, a dramatic departure from the current infrastructure. However, solutions to the most daunting energy issues cannot wait for a fossil-free energy future. There is a critical need for an energy supply and utilization infrastructure based on currently available processes that provides for current needs while greatly reducing the environmental consequences to future generations [1]. The global and multi-generational environmental concerns include traditional and climate-change pollutant emission, energy-driven water demand, resource depletion, and economics. It is clear that no one topic will solve all of the problems [2].

Fossil fuels provide about 85% of US and global energy, making it unrealistic to imagine a near-term, fossil-free infrastructure. Fossil fuels also dominantly contribute to global climate change. Therefore, carbon capture and storage (CCS) looms large in any transitional energy infrastructure. A recently patented process for CO_2 removal from flue gas [3-5] portends a substantial decrease in both cost and energy consumption compared to solvent- and oxygen-based systems. Some of the detailed issues associated with this technology include (a) characterization (size distribution, morphology) of desublimated CO_2 particles, (b) particle separation from gases and subsequent pressurization of the condensed phase at cryogenic temperatures, (c) process and scientific models of CO_2 removal as a function of operating conditions (temperature, pressure, composition) addressing CO_2 and trace gases (SO_2, NO, Hg, etc.), (d) standardized comparisons of CO_2 removal processes [6-8], and (e) providing laboratory-scale verification and validation of the predicted performance [6, 7, 9]. In these ways it is similar to the most thoroughly analyzed alternatives and some of their primary challenges, namely oxyfuel firing (burnout, deposit formation, heat transfer patterns) and combustor, gasifier, and pyrolyzer cofiring with biomass, including methods of increasing gasification and combustion efficiency and reliability [10-21].

Cryogenic Carbon Capture

The cryogenic carbon capture (CCC) process is a retrofit, post-combustion technology that desublimates CO_2 in the flue gas, separates the resulting solid from the remaining light gases, pressurizes the solid CO_2, melts the CO_2 and warms the light gas, and completes the CO_2 pressurization with the liquid CO_2. Desublimation occurs in one of several ways, two of which are discussed here. The first version (Figure 1) (1) dries and cools flue gas, (2) compresses it to 5-7 bar, (3) cools it in a heat recovery heat exchanger to nominally -107 ℃, (4) may extract condensed contaminants such as mercury, SO_2, NO_2, and HCl at various stages during cooling and condenses most (about 75%) of the CO_2 during cooling (cooling happens in stages even though it is shown as a single step in the figure), (5) expands the remaining light gas to further cool it to between -120 ℃ (90 % capture) and -135 ℃ (99 % capture), (6) separates the remaining solid CO_2 that forms during cooling from the remaining gas, (7) pressurizes the solid CO_2 to 70-80 bar, (8) reheats the CO_2 and the remaining flue gas to near ambient conditions (15 ℃) by cooling the incoming gases, and (9) compresses the pressurized and now melted CO_2 stream to final delivery pressure (nominally 150 bar). There is a small external refrigeration loop in the process that transfers the enthalpy of pure CO_2 melting to cooler temperatures to avoid a heat exchanger temperature cross over.

An alternate version of the process (Figure 2) passes through the same temperature points but uses an external refrigeration loop, which avoids compressing flue gas and passing it through a turbine. This increases the total system fluid flow and hence equipment size but resolves several issues associated with compression and expansion of flue gas. Most of the discussion that follows pertains to both systems except where noted.

A simplified flow diagram for the first version of the system using Aspen Plus modeling elements appears as Figure 3. Table 1 summarizes the stream names and their descriptions. Streams beginning with the letter "I" are internal streams that are not accurately simulated by Aspen because of solids formation limitations and are not included in the stream table due to size constraints. The warning tag on the process element near the SPUMP indicates that Aspen does not know how to characterize the performance of a pump for solids.

Gaseous and liquid compositions in streams appear separately from the solids compositions in Table 2, with the solid flows near the bottom of the table. Stream properties appear in several units. The species CO_2(s) differs from CO_2 in that the former is a solid and the latter shifts from a gas to a liquid, depending on local conditions. Recall that all stream names beginning with the letter I represent internal streams with inaccurate thermodynamic representations because of simulator limitations. These thermodynamics are corrected by the equilibrium reactors, and the compositions, mass balances, and energy balances of all other streams and the system as a whole are correct, with the exception that the heat exchanger requires additional energy input because of issues discussed later.

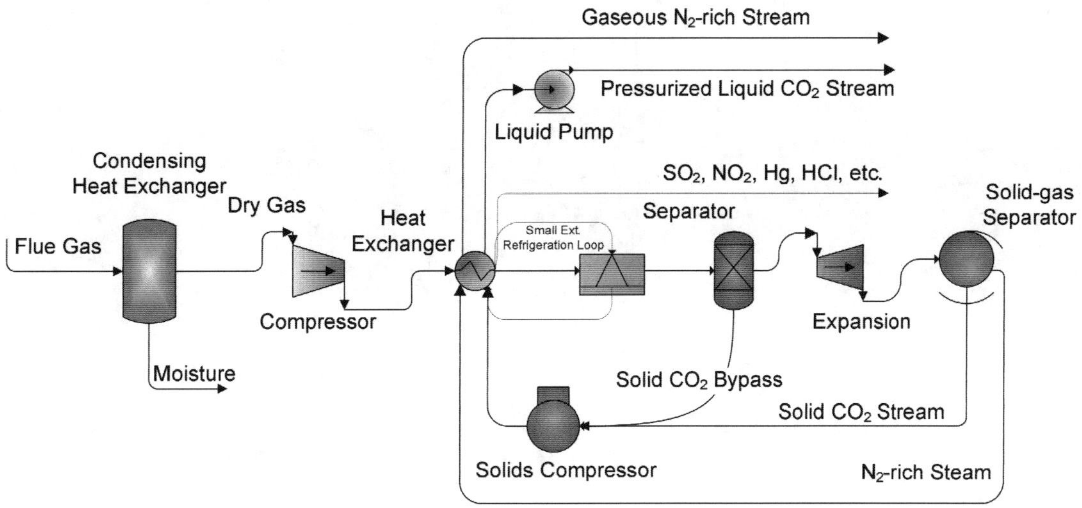

Figure 1. Simple schematic diagram of the cryogenic carbon capture (CCC) process that uses the flue gas as its own coolant.

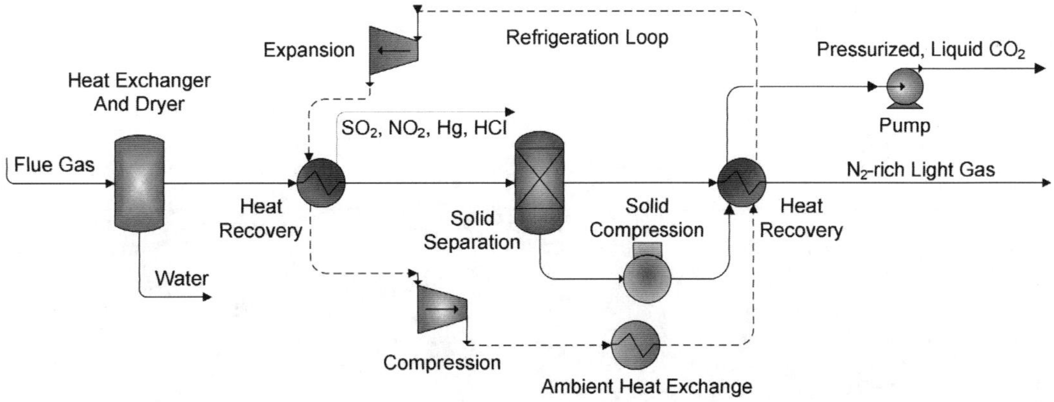

Figure 2. External, closed-loop heat exchanger version of the CCC process. Dashed line represents the closed-loop, external refrigeration cycle.

The heat exchanger represents one of the primary design challenges in the CCC process. There are several principal issues in the heat exchanger operation, including, in approximate order of decreasing difficulty: (1) preventing temperature profile crossing, that is, maintaining a small, consistent, and positive temperature difference between the warm and cool streams in the counter-current heat exchangers, (2) preventing frosting of heat exchange surfaces by desublimating CO_2, (3) designing a heat exchanger that potentially includes solids, gases, and liquids in different regions of the system, (4) potential corrosion and similar materials issues, and (5) managing safety and mechanical issues associated with modest (< 10 atm) and high (>50 atm) gas and condensed phase pressures, respectively, in the system. Much of the work in this document revolves around these issues.

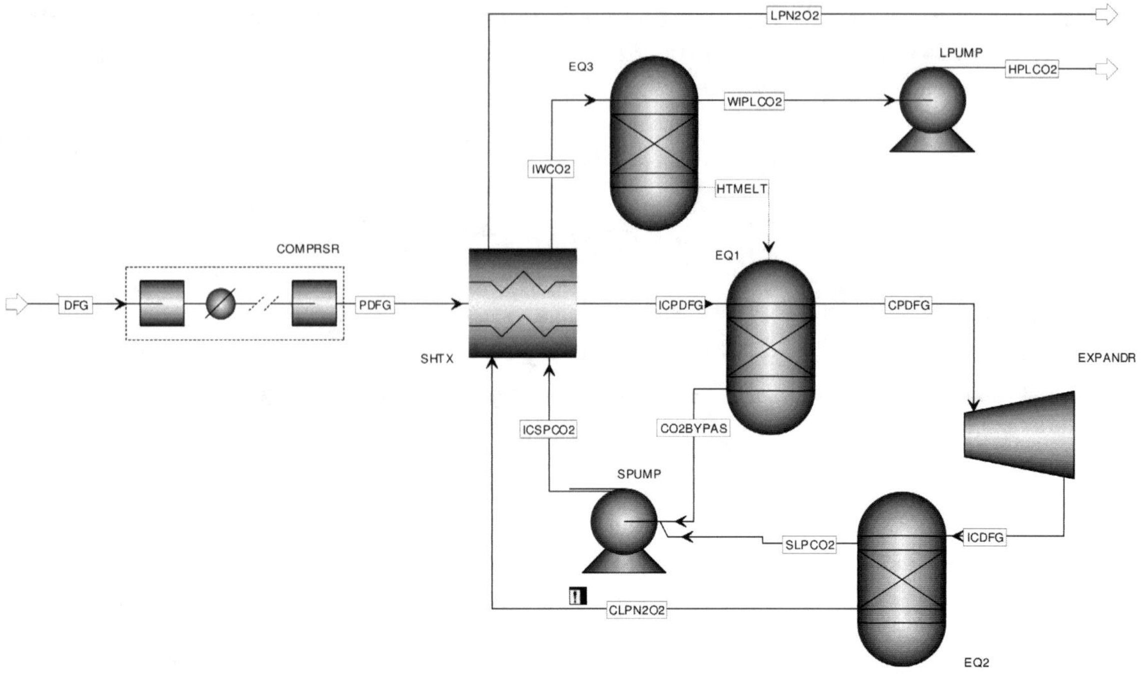

Figure 3 Aspen Plus flow diagram approximating the CCC system. The cluster of process elements SHTX, EQ1, and EQ3 approximate the multi-stream heat exchanger with solids freezing and melting and the cluster EXPANDR and EQ2 represents an expansion turbine with solids formation. The remaining elements are well represented without clusters.

Table 1 Stream IDs, descriptions, phases, nominal pressures, and nominal temperatures. Pressures and temperatures vary with desired capture efficiency. For streams with IDs that begin with I, the proper condensed phase appears in parentheses while the computed one appears outside the parentheses.

Stream	Description	Phase	Pressure	Temperature
DFG	Dry Flue Gas	G	1 bar	25 ℃
PDFG	Pressurized Dry Flue Gas	G	5 - 10 bar	25 ℃
ICPDFG	Internal Cold Pressurized Dry Flue Gas	G & L (&S)	5 - 10 bar	-95 ℃
CPDFG	Cold Pressurized Dry Flue Gas	G	5 - 10 bar	-95 ℃
CO2BYPAS	CO_2 Bypass	S	5 - 10 bar	-95 ℃
ICDFG	Internal Cold Dry Flue Gas	G & L (&S)	5 - 10 bar	-95 ℃
SLPCO2	Solid Low-pressure CO_2	S	1 bar	-120 to -135 ℃
CLPLG	Cold Low-pressure Light Gas	G	1 bar	-120 to -135 ℃
CSPCO2	Cold Solid Pressurized CO_2	S	70 bar	-120 to -135 ℃
LPLG	Low-pressure Light Gas	G	1 bar	-120 to -135 ℃
IWCO2	Internal Warm CO_2	S (L)	70 bar	25 ℃
WIPLCO2	Warm Intermediate-pressure Liquid CO_2	L	70 bar	25 ℃
HPLCO2	High-pressure CO_2	L	150 bar	25 ℃

Table 2 Typical flow stream properties for the CCC process operating at the conditions of

	DFG	PDFG	CPDFG	CO2BYPAS	SLPCO2	CLPLG	CSPCO2	LPLG	WIPLCO2	HPLCO2
Into	COMPRSR	SHTX	EXPANDR	SPUMP	SPUMP	SHTX	SHTX		LPUMP	
From		COMPRSR	EQ1	EQ1	EQ2	EQ2	SPUMP	SHTX	EQ3	LPUMP
State	VAPOR	VAPOR	VAPOR	MISSING	MISSING	VAPOR	MISSING	VAPOR	LIQUID	LIQUID
Substream: MIXED										
Mole Flow kmol/hr										
CO2	16044	16044	3618	0	0	90	0	90	15954	15954
CO2(S)	0.00	0.00	0.00	0.00	0.00	0.00	0.00	0.00	0.00	0.00
N2	99232	99232	99232	0	0	99232	0	99232	0	0
O2	3565	3565	3565	0	0	3565	0	3565	0	0
Total Flow kmol/hr	118841	118841	106416	0	0	102887	0	102887	15954	15954
Total Flow kg/hr	3600000	3600000	3053160	0	0	2897880	0	2897880	702122	702122
Total Flow cum/hr	2943280	443733	229424	0	0	1155240	0	2488470	850	987
Temperature C	25.00	25.00	-96.63			-138.20		15.00	15.00	25.61
Pressure bar	1.00	6.60	6.60			0.99	70.00	0.99	69.99	150.00
Vapor Frac	1.00	1.00	1.00			1.00		1.00	0.00	0.00
Enthalpy kcal/mol	-12.69	-12.71	-4.08			-1.22		-0.15	-96.80	-96.69
Entropy cal/mol-K	1.16	-2.63	-6.91			-5.19		0.12	-16.23	-16.21
Density kg/cum	1.22	8.11	13.31			2.51		1.16	825.92	711.64
Average MW	30.29	30.29	28.69			28.17		28.17	44.01	44.01
Substream: CISOLID										
Mole Flow kmol/hr										
CO2(S)	0.00	0.00	0.00	12425	3528	0.00	15954	0.00	0.00	0.00
Total Flow kg/hr	0.00	0.00	0.00	546839	155283	0.00	702122	0.00	0.00	0.00
Temperature C				-97	-138		-105			
Pressure bar	1.00	6.60		6.60	0.99		70.00	0.99		150.00
Density kg/cum				1591	1629		1602			

Performance Advantages, Development Status

The capture efficiency depends on the operating pressures of the compressor and success of heat integration steps. Table 3 summarizes representative results for the first version of the process assuming a best-of-class compressor (92% polytropic efficiency) and a dry, simple flue gas (Simple System column) and for a complex flue gas containing SO_x, NO_x, Hg, HCl, and saturated in moisture using a system with a supplemental refrigerator to address CO_2 melting issues (discussed later). Also, the simulation that produced these results assumes an equilibrium reactor, which converts some pollutants to forms that may not actually occur in a realistic system. Despite these simulation shortcomings, the table provides a reasonable measure of system performance.

Table 3 Figures of merit for the CCC process under simple (only N_2, O_2, and CO_2) and complex (flue gas containing S, N, Hg, and Cl impurities) assumptions.

Variable	Units	Simple System	Complex System
CO_2 In	kg/hr	706073	730057 (13.5%)
CO_2 Captured	kg/hr	702122	657108
Compression Energy	kW	175626	160300
Expander Energy	kW	-58340	-57444
Pump Energy	kW	1988	1486
Supplemental Refrig Energy	kW		17600
Specific Energy	GJ/tonne	0.601	0.620
CO_2 Capture Efficiency	-	0.995	0.90
SO_x Capture Efficiency	-		1.0000
NO_x Capture Efficiency	-		1.0000
Cl_2/HCl Capture Efficiency	-		0.001
Hg Capture Efficiency	-		1.0000
Usable H_2O Recovery	-		0.91
CO_2 Purity in Captured Stream	-	1.0000	1.0000

Figure 4 summarizes the variation of the specific energy consumption with capture efficiency and compression efficiency – two prominent parameters – assuming an average 10 °C temperature difference in the heat exchangers and typical specifications for the flue gas composition (13.5% CO_2 on a dry basis), pump and expander performance, etc. Significantly, CCC is capable of very high capture efficiencies without large increases in energy demand. For example, capture efficiencies of 99.7% are achieved when the coolest temperature is -143 °C, and these correspond to capturing all of the CO_2 produced from coal combustion and a fraction of that in the incoming air, where the nominal concentration of CO_2 in the air is 380 ppm. As indicated, this can be achieved with relatively modest increase in the specific energy requirements for the process with CCC.

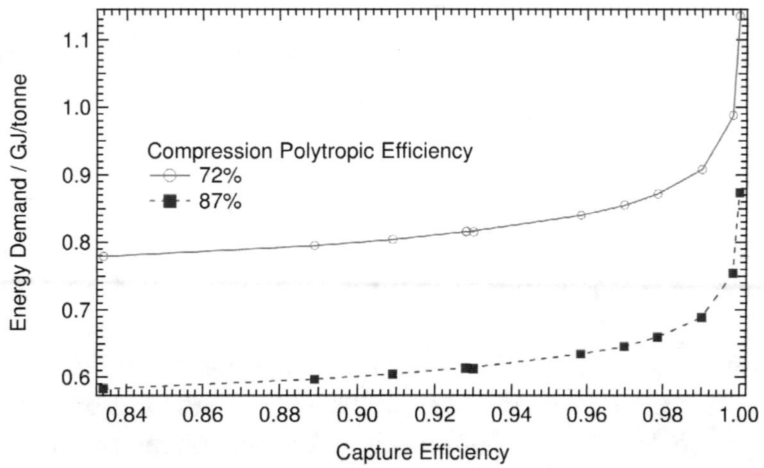

Figure 4. Variation in work-equivalent specific energy (GJ/metric ton of CO_2) demand of the CCC process as a function of capture efficiency and compressor efficiency.

The cost and energy efficiency of this process are its most compelling advantages. Several additional major advantages to this process include:

- Bolt-on retrofit to existing processes with minimal upstream modification.
- Simultaneous pollutant removal eliminates or reduces the need for other pollution control processes.
- Reduces power plant water demand, recovering up to 90% of flue gas water in a useful form.
- Eliminates toxic and hazardous streams and uses equipment and processes common to power generation.

Forms of energy used in alternative processes can include mostly work (oxy-fuel and gasification processes) or a combination of work and heat (solvent absorption processes). This complicates their direct comparisons. Conversion of work to heat or primary energy requires process-specific parameters, specifically the cycle efficiency of producing work. On a primary energy basis, the specific energy of the CCC process is about 1.62 GJ/tonne of captured CO_2. This, however, is less easily compared to other systems than it may seem, especially in an international context, since many countries assume a lower rather than higher heating value basis when computing the primary energy input to a process. Alternatively, the energy required by other processes can be converted to work-equivalent units, producing a number less susceptible to differences in international norms and that clearly reflects the impact on a power plant. Oxycombustion and amine absorption processes require about 1.8 and 2.2 GJ/tonne CO_2 captured, respectively, at 90 % capture efficiency in units of work when including lost capacity and impact on other plant processes, based on DOE simulations of these processes using the same software and very similar assumptions as we use here [6, 7, 22]. The equivalent work requirement of the CCC process at 90% capture efficiency is 0.53 GJ/tonne in the simulation shown in Figure 3, assuming a 92% efficient compressor. Under best-of-class conditions (92% efficient compressor, 5 ℃ average temperature difference in heat exchanger, 95% efficient turbine, etc.), the CCC process achieves 90% capture efficiency with nominally pure CO_2 as product at a specific energy cost of 0.49 GJ/ton CO_2 and with SO_x and NO_x effluent concentrations less than 1 ppm each (representing > 99% capture of these pollutants), approaching ¼ of the energy requirement of the alternatives. However, these conditions will prove difficult to achieve in an actual process. For example, heat exchanger design details increase this number, but it remains well below 50% of the energy requirements of the alternative processes. Previous publications and presentations by SES [23, 24] deliberately understate the potential performance advantages by a large fraction, in part because even the advantages discussed in these publications are sufficiently motivating to pursue this technology aggressively and in part because the resolution of some of the details of the process will inevitably compromise some of these best-of-class performance metrics. Additionally, some recent estimates indicate less energy impact for the solvent absorption process in particular, but still well above those for CCC. A quantitative comparison of this technology to amine absorption, the leading alternative technology, appears in Table 4.

Table 4 Comparison of several figures of merit between amine absorption and CCC processes, with CCC validation status. All results assume 90% capture and product CO_2 pressurization to 150 bar. Amine absorption data are from DOE estimates using the same software and approach as our estimates for CCC [5-7]. American Air Liquide and GE Global Research analyzed the process independent of SES and of each other and confirmed the cost and energy estimates. They and Dong Energy (Europe) have reviewed the process details and confirmed feasibility and qualitative advantages.

	Absorption	CCC	CCC Validation Status
Energy Demand (work equivalent, GJ/tonne CO_2 captured)	1.7-2.2	0.5-0.8	Model estimates, confirmed independently by 2 large commercial vendors & reviewed internationally
CO_2 Purity	High	Pure	Model and Preliminary Lab Data
SO_x Emissions	negligible	negligible	Model and Preliminary Lab Data
NO_x Emissions, lb/MMBtu	0.07	negligible	Model and Preliminary Lab Data
Hg Emissions	1 ppm	< 1 ppb	Model estimates
Inc. Energy for High CO_2 Capture	High	Modest	Model estimates
Retrofit Impact on System	Moderate	Minimal	Model estimates
Water Demand	High	Low	Model estimates and Preliminary Lab Data
Hazardous/Toxic Materials	High	Minimal	Model estimates and system demands
Increase in LCOE	> 80%	< 35%	Model estimates, confirmed independently by 2 large commercial vendors & reviewed internationally

Expander (Process version 1 only)

This discussion pertains only to the first process version. The second process version does not contain an expander in the flue gas stream, specifically to avoid the issues discussed here and with respect to the solids formation, etc. The second version does include a conceptually similar expander in the refrigeration cycle, but it operates on the refrigerant fluid, not the flue gas.

After the first equilibrium reactor (EQ1), the cold, solid CO_2 that formed in the heat exchanger bypasses the expander (CO2BYPAS) to minimize the amount of solid that the expander must handle. The gaseous remainder of the pressurized flue gas (CPDFG) enters an expander, where it performs work, further cooling the gas and condensing more CO_2. Similar to the heat exchanger, the process simulator does not allow solids to form in expanders/turbines. Instead, it predicts liquid formation. Therefore, the predicted temperature and physical state of the fluid at the expander exit are not correct and are hence labeled an internal cold dry flue gas (ICDFG). Another equilibrium reactor (EQ2) brings the stream to its proper equilibrium state, and the combination of the expander (EXPANDR) and the equilibrium reactor (EQ2) accurately describes the mass and energy balances of a real expander under these conditions. At this point, the gas temperature ranges from -120 °C to -135 °C, corresponding to 90% CO_2 capture and a compression ratio of about 5 and to 99% CO_2 capture and a compression ratio of about 6.8, respectively. Repeated laboratory experimental measurements using both simulated and actual coal flue gas and using both no expander and simulated expanders (gas expansion through nozzle with external cooling) have verified these capture efficiencies (Figure 5).

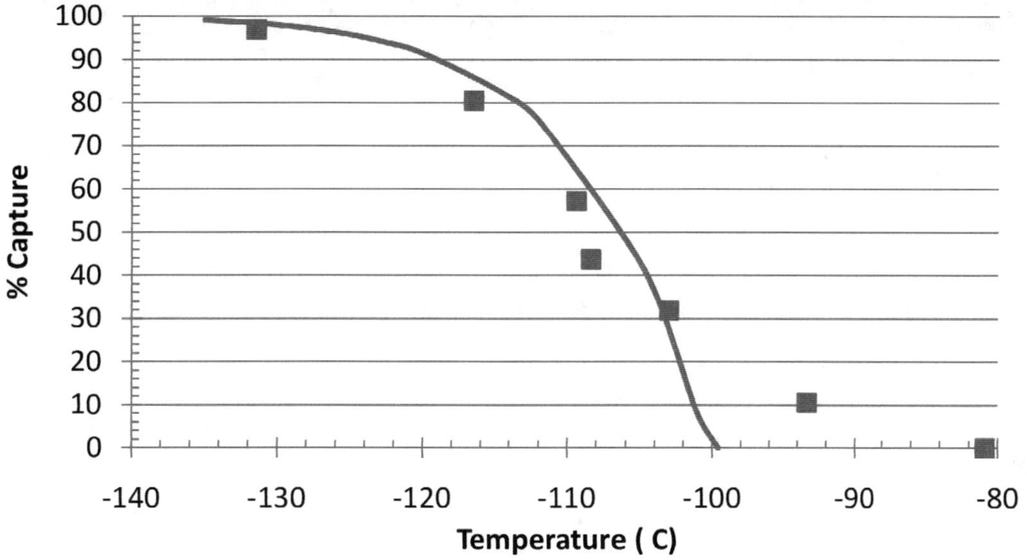

Figure 5 Experimental capture efficiency data as a function of temperature from laboratory experiments confirming predicted results.

Solids formation in a turbine represents a major concern. While the amount of solids formation typically ranges from 3 to 5 % and is well within the amount of liquid formation tolerable in most turbines, the fact that it is a solid is a concern for at least two reasons: erosion and deposition. The potential erosion of expander blades caused by entrained CO_2 particles represents a significant concern. Our recent research included an investigation of this potential. Using ASTM standard test procedures, CO_2 particles formed by flue gas expansion impacted on metal coupons representing different possible turbine materials produced no measureable erosion (Figure 6). Based on before (left) and after (right) comparisons of metal coupon surface features, no measureable erosion was observed. The discoloration of the coupon on the right is ink used to mark the impact point. The scratches are the natural metal surface features and do not change during testing. Continued testing of this type is being explored in the laboratory.

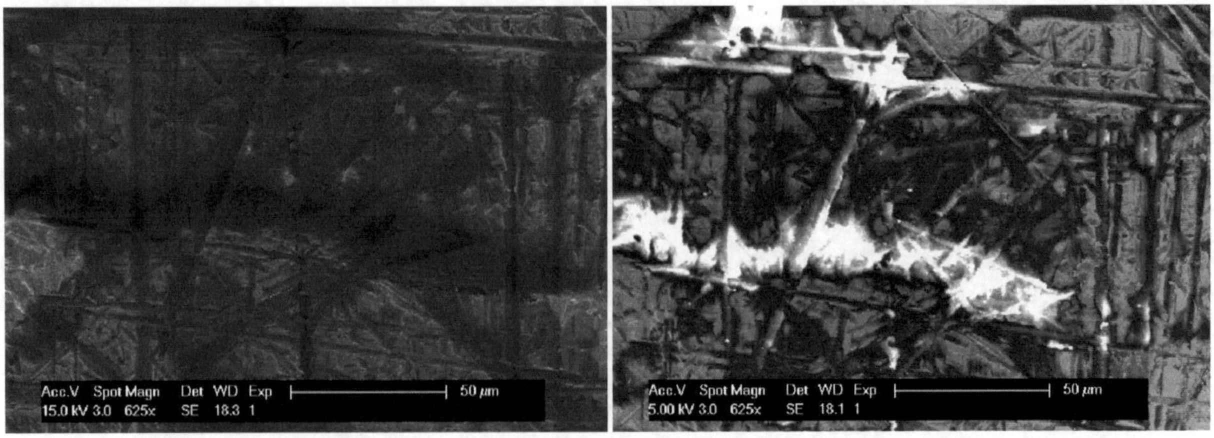

Figure 6 Laboratory data testing erosion potential of entrained CO_2 particles in an expander based on before (left) and after (right) comparisons of metal coupon surface features during particle impaction tests. No measurable erosion was measured. The discoloration of the coupon on the right is ink used to mark impact point. The scratches are the natural metal surface features and do not change during testing.

Both prior to and after the expander (in process version 1), particles formed in suspension must be separated from the light gases. Our recent laboratory experiments provide an indication of the particle size distributions formed during expansion and during cooling without expansion of simulated and real flue gases. In situ measurements of particle size at the exit of laboratory-scale heat exchangers and expansion coolers indicate that the particle sizes are approximately 600-750 microns (Figure 7). These are much larger than anticipated, but our experience has consistently been that fundamental predictions of aerosol sizes in industrial environments wildly underestimates the actual sizes [25]. Considerable care went into assuring that moisture and CO_2 condensation on surfaces did not affect these results. Nevertheless, these experiments and the accompanying theory remain under investigation. In particular, the concentration of CO_2 in the gas phase, which is quite low after the first separation of solids in process version 1, should affect these results.

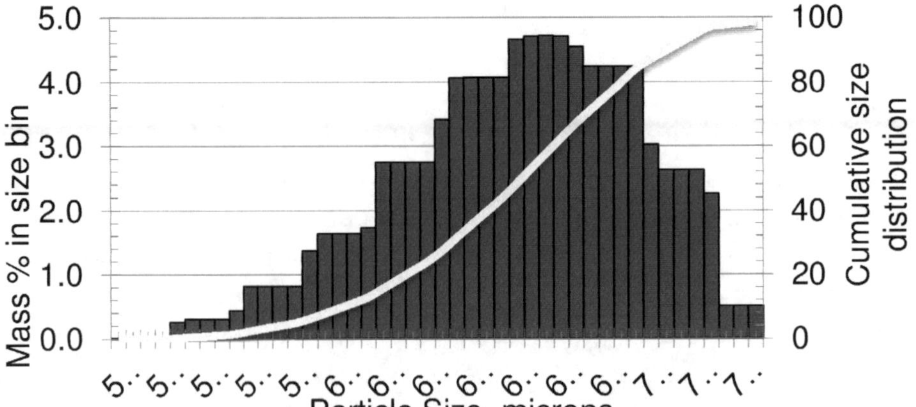

Figure 7 Measured in situ CO_2 particle size distributions typical of desublimation from simulated flue gases in both expansion and cooling tests.

Conclusion

Cryogenic separation of CO_2 from flue gases represents a cost effective and energy efficient alternative to providing a pure and highly efficient separation of CO_2 from flue gas. The process advantages stem from (in approximate order of importance) (a) elimination of energy- and entropy-intensive cyclic separation processes such as distillation and absorption, (b) compressing a condensed phase rather than a gas, (c) somewhat smaller dry volumetric flow rates, and (d) several ancillary process advantages (bolt-on technology, energy and water savings, simultaneous treatment of pollutants). Most of the essential aspects of the process have been demonstrated at lab or bench scale with positive results.

Bibliography

1. United Nations, *Report of the World Commission on Environment and Development 1987: Our Common Future*, G.H. Brundtland, Editor. 1987, United Nations: New York City.
2. Pacala, S. and R. Socolow, Stabilization wedges: Solving the climate problem for the next 50 years with current technologies. Science, 2004. **305**(5686): p. 968-972.
3. Baxter, L., *Carbon Dioxide Capture from Flue Gas*, in *US Patent Office*. 2008, Brigham Young University: US.
4. Baxter, L.L., *Simultaneous CO2 Capture and Energy Storage*, in *US Patent Office*. 2008, Brigham Young University: USA.
5. Baxter, L.L., The Application of Light Gas Turbines to Enhance the Energy Efficiency of Existing Steam Cycles in Power Plants, in US Patent Office. 2008, Brigham Young University: USA.
6. US Department of Energy, N.E.T.L., Cost and Performance Baseline for Fossil Energy Plants Volume 1: Bituminous Coal and Natural Gas to Electricity 2007.
7. US Department of Energy, N.E.T.L., Pulverized Coal Oxycombustion Power Plants Volume 1: Bituminous Coal to Electricity 2007.
8. Klara, J.M., *Fossil Energy Power Plant Desk Reference*. 2007, National Energy Technology Laboratory: Pittsburg, PA.
9. Wall, T.F., *Combustion processes for carbon capture.* Proceedings of the Combustion Institute, 2007. **31**: p. 31-47.
10. Baxter, L., Biomass-coal co-combustion: Opportunity for affordable renewable energy. Fuel, 2005. **84**(10): p. 1295-1302.
11. Baxter, L. and J. Koppejan, *Co-combustion of biomass and coal.* Euroheat and Power (English Edition), 2004(1): p. 34-39.
12. Baxter, L.L., Ash Deposition During Biomass and Coal Combustion: A Mechanistic Approach. Biomass and Bioenergy, 1993. **4**(2): p. 85-102.
13. Jenkins, B.M., et al., *Combustion properties of biomass.* Fuel Processing Technology, 1998. **54**(1-3): p. 17-46.
14. Kaufmann, H., et al., Deposit formation on a single cylinder during combustion of herbaceous biomass. Fuel, 2000. **79**: p. 141-151.
15. Lokare, S.S., et al., Investigation of ash deposition rates for a suite of biomass fuels and fuel blends. Energy & Fuels, 2006. **20**(3): p. 1008-1014.

16. Lu, H., et al., Effects of particle shape and size on devolatilization of biomass particle. Fuel, 2009.
17. Lu, H., et al., *Comprehensive study of biomass particle combustion.* Energy & Fuels, 2008. **22**(4): p. 2826-2839.
18. Miles, T.R., et al., *Boiler deposits from firing biomass fuels.* Biomass and Bioenergy, 1996. **10**(2-3): p. 125-138.
19. Nielsen, H.P., et al., *Implications of chlorine-associated corrosion on the operation of biomass-fired boilers.* Progress in Energy and Combustion Science, 2000. **26**: p. 283-298.
20. Robinson, A.L., H. Junker, and L.L. Baxter, *Pilot-scale investigation of the influence of coal-biomass cofiring on ash deposition.* Energy & Fuels, 2002. **16**(2): p. 343-355.
21. Robinson, A.L., et al., *Interactions between coal and biomass when cofiring.* Symposium (International) on Combustion Proceedings of the 1998 27th International Symposium on Combustion,Aug 2-Aug 7 1998, 1998. **1**: p. 1351-1359.
22. US Department of Energy, N.E.T.L., Pulverized Coal Oxycombustion Power Plants Volume 1: Bituminous Coal to Electricity: Revision 2. 2008.
23. Baxter, L.L., A.P. Baxter, and S.S. Burt, Cryogenic CO2 Capture as a Cost-Effective CO_2 Capture Process, in International Pittsburgh Coal Conference. 2009: Pittsburgh, PA.
24. Baxter, L.L., A.P. Baxter, and S.S. Burt, Cryogenic CO_2 Capture to Control Climate Change Emissions, in Coal: World Energy Security - The Clearwater Clean Coal Conference - The 34th International Technical Conference on Clean Coal & Fuel Systems. 2009: Clearwater, Florida.
25. Baxter, L.L., et al. Particle Size Distributions in Recovery Boilers. in International Chemical Recovery Conference. 2001. Whistler, British Columbia, Canada: TAPPI.

(127b) Characteristics of Aminosilicones Used for CO2 Capture

Sarah Genovese[1], Benjamin R. Wood[1], Robert J. Perry[1], Michael O'Brien[1] and Sam D. Draper[2], (1) Global Research, General Electric, Niskayuna, NY, (2)GE Energy, Greenville, SC

This work describes the characterization of the material and CO_2 absorption characteristics of a new class of CO_2 absorption liquids. Mixtures of aminosilicones and hydroxyethers were shown to have improved dynamic CO_2 capture capacities compared to more common CO_2 capture solvents such as 30% MEA. Isotherms and other measured properties were used to calculate the cost of electricity using a CO_2 capture system model integrated into a pulverized coal-fired power plant system. Systems for study in a lab-scale continuous absorption / desorption apparatus were based on these results.

See more of this Session: Unconventional Technologies for CO2 Capture, Conversion and Utilization I
See more of this Group/Topical: Topical G: Innovations of Green Process Engineering for Sustainable Energy and Environment

(127c) Post-Combustion CO2 Capture Via Formation of Aminosilicone Carbamate Salt Powders in Spray Dryers

Robert M. Enick[1], Karl Johnson[2], Hong-bin Xie[3], Deepak Tapriyal[4], **Bing Wei**[5], Robert J. Perry[6], Sarah Genovese[6], Benjamin R. Wood[6] and Michael O'Brien[6], (1)Department of Chemical and petroleum Engineering, University of Pittsburgh, Pittsburgh, PA, (2)University of Pittsburgh, Pittsburgh, PA, (3) Dept.of Chemical and Petroleum Engineering, University of Pittsburgh, Pittsburgh, PA, (4)Chemical and Petroleum Engineering, University of Pittsburgh, Pittsburgh, PA, (5)Chemical and Petroleum Engineering Department, University of Pittsburgh, Pittsburgh, PA, (6)Global Research, General Electric, Niskayuna, NY

ABSTRACT BODY: A particular low viscosity aminosilicone compound terminated with a primary amine on each end (designated as GAP-0) has been identified that reacts very quickly with gaseous CO2 to form a white solid carbamate salt. If the stoichiometric amount of CO2 reacts with GAP-0, the ratio of the mass of CO2 in the GAP-0 carbamate salt to the mass of the GAP-0 is 18%.

Therefore, pure liquid GAP-0 was displaced through the nozzle of a lab-scale spray dryer, where the resultant droplets reacted with CO2 gas flowing co-currently through the spray chamber. The carbamate salt powder formed in the chamber was separated from the CO2-depleted gas stream in a cyclone separator. (Note the while a spray "dryer" apparatus was used, the formation of the solid powder is actually due to the reaction between the GAP-0 liquid and CO2 gas; it is not due the evaporation of a liquid.)

We propose that this process may serve as a novel means of CO2 capture that eliminates the need for a solvent. Therefore it is likely to reduce regeneration heat requirements because there would be no requirements for the sensible and latent heat demands associated with an aqueous or non-aqueous solvent. The carbamate salt powder would be collected from the bottom of the cyclone separator and then heated to release the CO2, allowing the GAP-0 liquid to be reclaimed. The GAP-0 would then be recycled to the nozzle at the top of the spray chamber, while the CO2 would be compressed for sequestion.

The properties of the carbamate salt powder product were determined, including particle size (SEM), melting point (DSC), heat required for regeneration (DSC), phase behavior and rate of CO2 release as a function of temperature, pressure, time and ambient gas composition (variable volume PVT cell), and weight loss due to evaporation and CO2 release as a function of temperature in an inert argon atmosphere (TG MS), and in a CO2 atmosphere (TG).

(127d) Micro-Structured Adsorption Bed for Rapid CO2 Capture & Regeneration

Wei Liu[1], Haiying Wan[2], Shari Li[1] and Nathaan Canfield[3], (1)Energy and Environmental Directorate, Pacific Northwest National Lab, Richland, WA, (2)Pacific Northwest National Lab, Richalnd, WA, (3) Pacific Northwest National Lab, Richland, WA

Structured beds comprising a number of parallel flow channels at sizes of micrometer or sub micrometer level will be presented for capturing CO2 from a large volume of gas stream at short residence time and low energy cost. Such micro-structured beds are fabricated by confining fine adsorbent materials with thin, macro-porous barrier sheets (<200μm). The structure looks like the monolith catalytic converter but provides adsorbent loadings (>60 wt.%) much higher than the conventional washcoating (10 to 20 wt%). The overall mass and heat transfer resistance during adsorption and regeneration process can be minimized by independent designs of channel geometries and adsorbent layer structures (thickness, pore size, porosity). As a result, the adsorption and regeneration can be performed at a much faster cycle speed than the conventional pellet-loaded adsorbent bed. Key design principles will be delineated by experimental results with 13X and silicalite zeolite adsorbent materials. The structured bed preserves adsorption kinetics of the powder material and at the same time, reduces the pressure drop by 10 times compared to the 13X bead bed. Scope engineering design and analysis shows possibility to reduce power consumption by >70% and the adsorbent bed size by >5 times for capturing CO2 from flue gas streams by use of a structured adsorption bed, compared to the conventional bead bed structure. The structured bed decouples optimization of adsorbent material properties from the adsorption bed fabrication, and enables quick scale-up of the high-performance adsorbent material from powder form to the engineered adsorption bed without going through the pelletization process.

See more of this Session: Unconventional Technologies for CO2 Capture, Conversion and Utilization I
See more of this Group/Topical: Topical G: Innovations of Green Process Engineering for Sustainable Energy and Environment

(127e) CO2 & Steel Slag for Industrial Mineral Carbonation: Chemistry and Economics Are On Our Side

Michael Wyrsta, CarbonX Inc.,, Santa Barbara, CA

As the world rapidly industrializes, the environment pays a heavy cost; governments around the world are committed to tackling the problem of greenhouse gas emissions. There are roughly 6000 coal fired units in the world and growing. China and India are adding between 1-2 gigawatts of capacity every week. World market energy consumption will rise from 536 quadrillion KJ in 2010 to 715 quadrillion KJ in 2030. Almost 211 quadrillion KJ will come from burning coal. Likewise steel production will continue to grow. It is the dominant metal produced in the world, 95% of all metal produced is steel. Steel will also continue to be energy intensive and therefore a CO2 intensive industry emitting ~2 tons of CO2 for every ton of steel produced. The industrial mineral carbonation offers a ready opportunity to be one of the first globally deployable technologies to capture, quantify and sequester CO2. More importantly, mineralization technology offers a distinct advantage over geo-sequestration: positive revenue generation with a policy neutral position. We explore the economics and chemistry of using CO2 to convert steel slag into several product slates.

See more of this Session: Unconventional Technologies for CO2 Capture, Conversion and Utilization I
See more of this Group/Topical: Topical G: Innovations of Green Process Engineering for Sustainable Energy and Environment

(127f) Ultra-Stable Calcium Oxide Absorbents for High Temperature CO2 Capture

Liyu Li[1], David L. King[2], Zimin Nie[3], Xiaohong Li[2] and Chris Howard[1], (1)Hydrocarbon Processing Group, Pacific Northwest National Laboratory, Richland, WA, (2)Pacific Northwest National Laboratory, Richland, WA, (3)Pacific Northwest National Labratory, Richland, WA

Calcium oxide-based materials are attractive regenerable absorbents for separating CO2 from hot gas streams because of their high reactivity, high CO2 capacity, and low material cost. Especially, their high carbonation temperature (600oC to 700oC) makes it possible to recover and use the high quality heat released during CO2 capture, which cam largely increase the overall process efficiency. However, the performance of all reported CaO-based absorbents deteriorates as the number of carbonation-decarbonation cycles increases. Rapid loss of CO2 capacity over many carbonation/decarbonation cycles is always observed due to severe absorbent sintering. Recently, we have found that this sintering effect can be effectively mitigated by properly mixing with a modest amount of small particles of inert additives. A new class of CaO-based absorbents with much improved durability and CO2 reactivity were prepared by physical mixing of Ca(CH3COO)2 with MgAl2O4 spinel nanoparticles followed by high temperature calcination. A CaO-MgAl2O4 (32 wt.% spinel content) material provides 34 wt % CO2 capacity after 65 carbonation-decarbonation cycles (650oC and 850oC, respectively), corresponding to 63 % CaO utilization. Under the same test conditions, the CO2 capacity of natural dolomite (35 wt.% MgO and 65 wt.% CaO) decreases rapidly from 25 wt.% for the first cycle to less than 5 wt.% for the 50th cycle.

See more of this Session: Unconventional Technologies for CO2 Capture, Conversion and Utilization I
See more of this Group/Topical: Topical G: Innovations of Green Process Engineering for Sustainable Energy and Environment

(127g) Carbon Capture by Carbonate Mineralization

Anjana Meel, University of California, Santa Barbara, Santa barbara, CA and Michael F. Doherty, Department of Chemical Engineering, University of California at Santa Barbara, Santa Barbara, CA

It is now widely accepted that carbon dioxide (CO2) capture and sequestration has become one of the defining problems of our time. Of all the ways that have been suggested to achieve this, our approach in collaboration with Calera Corporation (based in Los Gatos, California) is to mimic nature's way and mineralize carbon dioxide into calcium carbonate (limestone) particles at the emission sites. In particular, post-combustion stack gases at atmospheric pressure are reacted with subsurface geological brines in industrial reactors to form solid carbonates (so called supplementary cementitious material or SCM) that are suitable for use in the construction industry. This concept has the potential benefit of synergistically integrating the power generation and cement manufacturing industries to make a useful product out of harmful CO2.

In this research, the two scenarios of the Calera technology for capturing 90% CO2 are investigated: Case A) All alkalinity is provided by brine, Case B) 15% of alkalinity is provided by an electrochemistry process. Initially, a deeper understanding of the chemistry is developed by identifying the total numbers of independent chemical equations to fully describe the process in the domain of given species (ions and molecules) and elements [1]. Subsequently, overall mass and energy balances are performed to estimate: (1) the quantities of sub-surface brines needed for capturing 90% CO2 and the subsequent amount of product produced (2) the energy penalty for the power plant while supporting the energy needs of the Calera carbon capture technology. Approximately 10% (energy penalty) of the total electric power produced is used by the process in Case A while 24% (energy penalty) of the total electric power produced is consumed in Case B. Note that the energy penalty for the liquid CO2 (i.e. CO2 separated from post-combustion stack gases in the form of a highly compressed and almost pure fluid) route of carbon capture and storage can range from 20% to 40% with an average number around 29% of the electric power generated from the power plant [2]. In addition, sensitivity analysis is carried out to identify feasible regimes of operation with respect to brine concentrations and dynamic depth, alkalinity contribution from electrochemistry, % CO2 capture, and % reaction conversion.

A deeper understanding of the mass transfer, reactions kinetics, and reaction rate laws, and thermodynamic behavior of CO2 absorption in to the brine solutions is currently underway and will be reported in the presentation. Sensitivity analyses are carried out to identify the process conditions that are favorable from a thermodynamic and kinetic point of view which will be utilized in designing the absorber. These will be used in finding process alternatives for the technology. For example, should absorption of CO2 into alkaline brine occur separately from reaction with hard brine (absorption followed by reaction & precipitation) or should it be performed simultaneously (absorption + reaction + precipitation).

References:

[1] Gadewar S. B., Doherty M. F., and Malone M. F., A systematic method for reaction invariants and mole balances for complex chemistries, Comp. Chem. Eng., 2001, pp 1199-1217

[2] House KZ, Harvey CF, Aziz MJ and Schrag DP, The energy penalty of post-combustion CO2 capture & storage and its implications for retrofitting the U.S. installed base, Energy Environ. Sci., 2009, 2, pp 193 – 205.

See more of this Session: Unconventional Technologies for CO2 Capture, Conversion and Utilization I
See more of this Group/Topical: Topical G: Innovations of Green Process Engineering for Sustainable Energy and Environment

(140a) Progress in Nutrient Recovery From Anaerobically Digested Wastes

Shulin Chen, Quanbao Zhao and Craig Frear, Biological Systems Engineering, Washington State University, Pullman, WA

Recovering energy from solid wastes such as organic fraction of municipal wastes and animal manures has been recognized as having great potential for generating renewable energy and reducing greenhouse gas emissions. A more desirable concept, however, is bio-refining with which the waste materials are converted to a series of co-products in addition to biogas. Anaerobic digestion is the most suitable technology for such an application because of its capability in handling mixed waste feedstocks at a relatively low cost. A major technical challenge is to identify targeted products and develop related processes to produce these products economically. Due to the fact that animal manure and many types of municipal wastes such as food wastes are high in nutrient contents, recovering these nutrients as co-products not only benefits resource recycling but also contributes to environmental protection in terms of air and water quality. The Bioprocessing and Bioproduct Engineering Laboratory at Washington State University has been actively investigating the chemistry of the effluents from dairy manure and food waste anaerobic digestion and from that chemical research, developing technologies for recovering nutrients from the effluents. This paper provides a summary of the most recent progress. The information will be useful to the engineering community in selecting and designing systems to recover nutrients from waste streams.

See more of this Session: Conversion of Solid Wastes to Energy and/or Product
See more of this Group/Topical: Topical G: Innovations of Green Process Engineering for Sustainable Energy and Environment

(140b) A Spectrophotometric Method for Quantitative Determination of Xylose In Fermentation Medium

Patrisha J. Pham[1], Rafael Hernandez[1], Benjamin Estill[2], Andro H. Mondala[1] and William T. French[1], (1)Dave C. Swalm School of Chemical Engineering, Mississippi State University, Mississippi State, MS, (2)Department of Biological Sciences, Mississippi State University, Mississippi State, MS

The characterization of released sugars during fermentation is very important. Difficulty arises in the availability of a rapid, inexpensive and sensitive method for the detection of target sugars because fermentation broths are a complex mix of nutrients, waste products and cell debris. Since many of the ingredients are non-chromophoric and can't be detected by absorbance, target sugars can be identified from the rest of the fermentation medium by its conversion to a chromophoric substance that can be measured spectrophotometrically. A method involving UV spectrophotometry for the quantitative determination of xylose as the target sugar was developed. This widely known method is based on the dehydration of xylose to furfural in acidic medium with heat. The furfural produced condenses with orcinol in the presence of Fe^{3+} as catalyst to form a colored complex that can be measured by UV spectrophotometry. Factors affecting xylose concentration measurements such as acidic medium concentration, heating time and amount of Fe^{3+} catalyst were investigated. A continuous scan revealed the working wavelength to be 671 nm. The effect of other components in the fermentation broth was found to be negligible. Absorbance shows a linear relationship to xylose in the range of 100-500 mg/L. The xylose concentrations from actual fermentation were determined with the method and compared with YSI 2700, an enzyme electrode employed for xylose measurements and HPLC-ELSD method, a common technique for measuring xylose. Dilution is necessary for comparable xylose concentrations with YSI 2700 and HPLC.

See more of this Session: Conversion of Solid Wastes to Energy and/or Product
See more of this Group/Topical: Topical G: Innovations of Green Process Engineering for Sustainable Energy and Environment

(140c) Transesterification of Activated Sludge Oil by Ionic Liquid Supported Acid Catalysis

Patrisha J. Pham[1], Rafael Hernandez[1], Dyane Marie Acevedo Vélez[2] and William T. French[1], (1)Dave C. Swalm School of Chemical Engineering, Mississippi State University, Mississippi State, MS, (2) Chemical Engineering, University of Puerto Rico - Mayagüez, Mayagüez, PR

Progress in the transesterification of vegetable oils has been achieved using ionic liquids. Ionic liquids have been found to provide an ideal medium for increasing the biodiesel yield, stabilizing the inorganic acid and removing the glycerol by-product. In this present study, transesterification of triglycerides common in activated sludge oil with methanol to biodiesel was investigated using various imidazolium-based ionic liquids under acidic conditions. The influences of the nature of the anion and the alkyl groups on the imidazolium cation, ionic liquid dosage, reaction temperature and reaction time were also explored.

See more of this Session: Conversion of Solid Wastes to Energy and/or Product
See more of this Group/Topical: Topical G: Innovations of Green Process Engineering for Sustainable Energy and Environment

(140d) Pressure Cycles of Ozonation and Aeration for Activated Sludge Solubilization

Chia-Jung Cheng and Andy Hong, Civil & Environ. Eng., University of Utah, Salt Lake City, UT

Waste activated sludge from wastewater treatment plants is a universal problem. Traditional disposal methods such as composting, land application, and incineration post environmental and cost issues. Anaerobic digestion has been most common for excess waste sludge; however, retention time is often extensive because of kinetic limitation in the hydrolysis step. Ongoing researches have shown considerable benefits of solubilizing sludge prior to anaerobic digestion in solids reduction as well as biogas formation.

We tested a new sludge solubilization technique to obtain increased yields of soluble cell materials. The new technique incorporates rapid, successive compression and decompression cycles using ozone and air only. During compression, steepened gas concentration gradient across the cell membrane drives dissolved gas into the cell, increasing the gas concentration within the cell. When the pressure is rapidly reduced, gas formation and expansion occur within the cell, causing the formation of gas pockets in the cell (as transport across the membrane is slower than decompression). Gas accumulation in the cells during repeated pressure cycles leads to eventual rupture of the cells. Microscopic pictures of activated sludge were taken prior to pressure cycles, after 10 pressure cycles with air reaching 1000 kPa, and after 10 pressure cycles with 0.06% O3 reaching 1000 kPa. Prior to treatment, an abundance of filamentous bacteria and floc along with discernable microbes was present as in typical activated sludge. Pressure cycles with air show similar but diluted biomass with ciliated stalks broken into smaller pieces, while pressure cycles with ozone reveal little biomass but disperse nondescript flocs and seemingly gluey patches in the liquid. The application of pressure cycles resulted in disintegration of activated sludge. In its final form, this paper will present solubilization results under different rupturing conditions as characterized by TS, TSS, TDS, VDS, VSS, sCOD, tCOD, BOD, as well as confirmatory identifications of DNA and protein in the slurry before and after treatment.

See more of this Session: Conversion of Solid Wastes to Energy and/or Product
See more of this Group/Topical: Topical G: Innovations of Green Process Engineering for Sustainable Energy and Environment

(140e) An Evaluation of Production of Fuels From Sewage Sludge

Y. Madueke, Ramalingam Subramaniam, Cunwen Wang, Stephen Dufreche, Mark Zappi and **Rakesh Bajpai**, Chemical Engineering, University of Louisiana at Lafayette, Lafayette, LA

Efforts to produce biofuels as alternatives for petrofuels often runs into issues of renewable raw materials tat can be utilized. As a result, we have explored use of sewage sludge for biofuels. This paper will discuss the different constituents in sewage sludge that can be used for this purpose. Lipids and polymers present in the sludge will be identified and chemistries for their transformations will be presented.

See more of this Session: Conversion of Solid Wastes to Energy and/or Product
See more of this Group/Topical: Topical G: Innovations of Green Process Engineering for Sustainable Energy and Environment

APPLICATION OF COAL FLY ASH AS A SOLID BASIC CATALYST IN PRODUCING BIODIESEL

Omotola Babajide[1,*], Leslie Petrik[1], Nicholas Musyoka[1], Bamikole Amigun[1,2], Farouk Ameer[1]

[1]Environmental and Nano Sciences group, University of the Western Cape, South Africa
[2]Sustainable Energy Futures, Council for Scientific and Industrial Research (CSIR) Stellenbosch, South Africa

Keywords: Fly ash, Heterogeneous catalyst, Sunflower oil, Transesterification, Methyl ester

Abstract

South Africa is largely dependent on the combustion of coal for electricity production; Eskom's coal-fired power stations consume approximately 109 million tons of coal per annum, producing around 25 million tons of ash, to supply the bulk (93%) of South Africa's electricity. The management of this fly ash has been a concern with various approaches for its beneficial use being investigated. This work presents the results of transesterification reaction using sunflower oil as feedstock with methanol and class F fly ash catalyst derived from a coal fly ash dump in South Africa to produce methyl esters (biodiesel). The influence of various reactions parameters such as loading amount of active components, methanol: oil ratio, reaction time, temperature and catalyst deactivation was investigated and the product synthesized was tested. The important fuel properties of the methyl esters compared well with ASTM biodiesel standard.

Introduction

The recent energy crisis, largely driven by factors of economic growth and social integration, has lead to the construction of new coal power plants in South Africa. Specifically, South Africa generates 93 % of its electricity from coal combustion. Eskom's coal-fired power stations consume approximately 109 million tons of coal per annum, producing around 25 million tons of ash, to supply the bulk of South Africa's electricity [1]. Almost 90% of the ash produced in the generation process is called fly ash or pulverised fuel ash (a waste by-product of burning coal to generate electricity). The disposal and management of the huge quantities of fly ash generated has been a concern to the policy advisers, industrial and the scientific communities. The utilization of fly ash as a suitable feedstock for use as heterogeneous catalyst for the transesterification would allow beneficiation of fly ash in an environmentally friendly way which would make the process more economically viable.

Biofuels are derived from renewable sources such as naturally occurring fats and oils, which may be obtained from variety of plants and animals. Since, diesel-fuel figures as an essential function in the industrial economy, biodiesel-synthesis has constituted the essence of current research in the field of bio-fuels [2]. Biodiesel, a green fuel, is comprised of mono-alkyl esters of long chain fatty acids which are derived from fatty acid glycerides, and are major components of an oil or fat. Transesterification is the key and foremost important process to convert vegetable oil or animal fat into a diesel-like fuel [3]. This is illustrated in Figure 1.

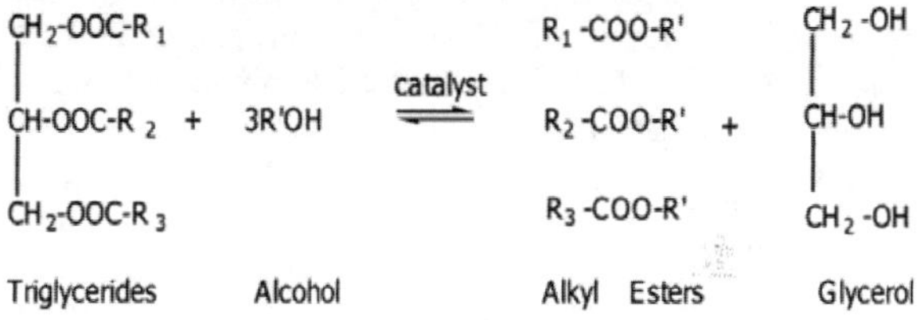

Fig. 1: Transesterification of Triglyceride with Alkyl Alcohol

While searching for promising solid base catalysts, several attempts have been made using basic zeolites, alkali and alkali earth oxides, alkali and alkali earth carbonates, supported guanidine and basic hydrotalcites. [4]. Fly ash, an inorganic residue arising from coal combustion processes consists essentially of SiO_2 and Al_2O_3 and being an environmental pollutant, constant generation of fly ash in huge quantities is a matter of concern. In view of its important economical and environmental implications, the disposal and utilization of fly ash represents a significant challenge. Development of environmentally safe applications for the utilization of coal fly ash is very important. Consequently, attempt was made to develop an efficient, inexpensive and environmental friendly fly ash-based heterogeneous catalyst for the esterification and/or transesterification reactions, for example for the production of biodiesel from sunflower oil.

Experimental

Table 1

Fatty Acid composition of Sunflower oil

Fatty Acid	Content (%)
Palmitic (16:0)	6.08
Stearic (18:0)	3.26
Oleic (18:1)	16.93
Linoleic (18:2)	71.73
Linolenic (18:3)	0.56

Pulverized fly ash collected from the ash collection systems at Arnot Thermal Power station in the Mpumalanga District of South Africa was used as a support. The elemental chemical composition of the fly ash (on dry basis) was SiO_2 50.91%, Al_2O_3 30.91%, Na_2O 0.10%, CaO 6.2%, Fe_2O_3 3.46%, MgO 1.48%, TiO_2 1.65%, MnO 0.02%, K_2O 0.60%, P_2O_5 0.56%. Before its use, material was ground to a fine powder. The average SiO_2/Al_2O_3 ratio of Arnot fly ash was found to be 1.65. The major elemental chemistry of the Arnot fly ash is consistent with $SiO_2+Al_2O_3+Fe_2O_3 \geq 70\%$ meaning that it is "class F" fly ash (ATSM method C 618). This is a typical composition of fly ash from combustion of bituminous coal from South Africa [5]. The alkali salt KNO_3 (99% purity from Merck Chemicals, South Africa) was used in as-received form to prepare its stock solution (aq. 5% wt/wt). Apart from fly ash from a pulverized coal power plant, another primary source of fly ash is from a municipal waste incineration plant.

Catalyst preparation and characterisation

The catalysts were prepared using established method of wet impregnation and were characterized to measure the properties such as pore size, surface area, structure to thus understand the performance of the catalyst using the following techniques. (X-ray powder diffraction (XRD), Scanning electron microscope (SEM), Fourier transformed infrared (FTIR) spectroscopy, BET surface area)

Results and discussion

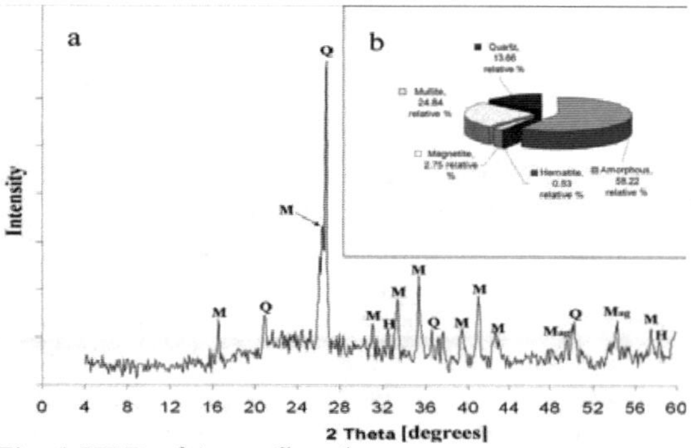

Fig. 1 XRD of Arnot fly ash

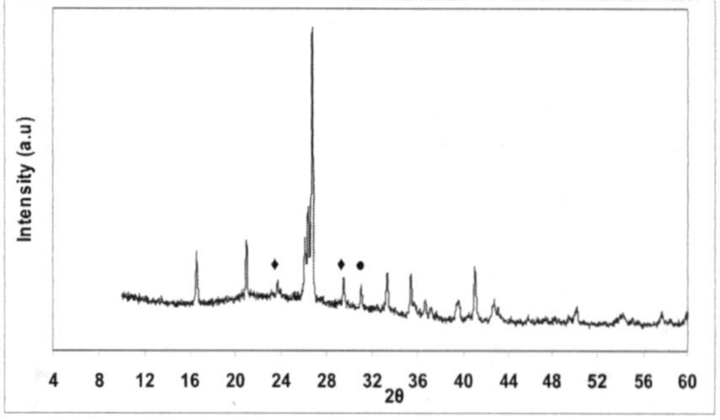

Fig. 2 XRD patterns of potassium impregnated fly ash catalyst with characteristic peaks due to KNO_3 • and K_2O♦.

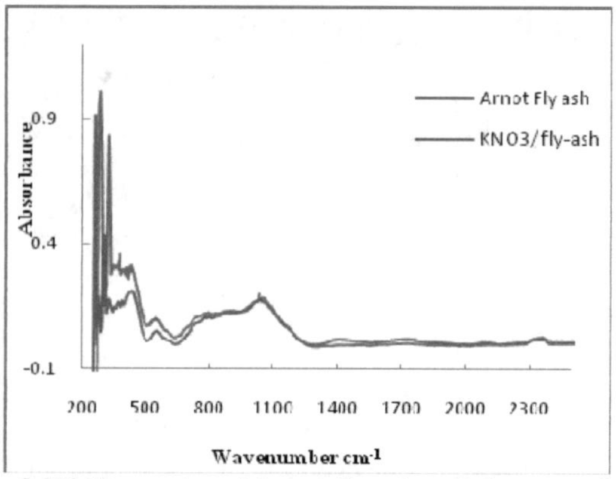

Fig. 3 FT-IR spectra of Arnot fly ash and the potassium impregnated fly ash catalyst

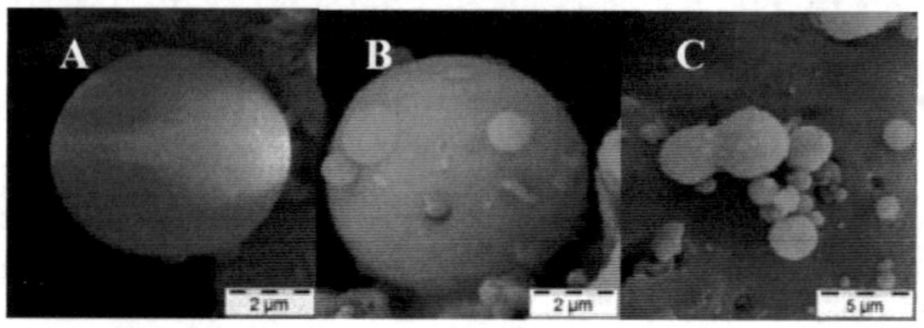

Fig. 4 SEM images of the (a) Arnot fly ash (b) and (c) potassium impregnated fly ash catalyst

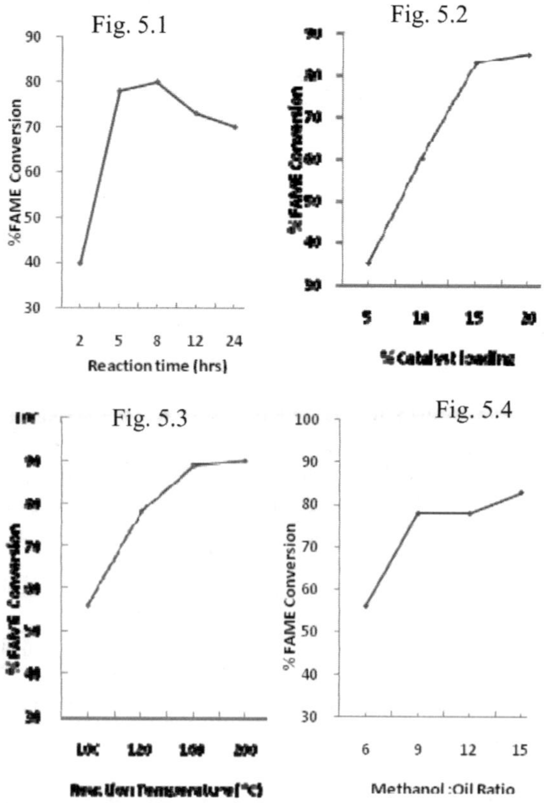

Figs. 5.1-5.4 Influence of process parameters on FAME conversion using the impregnated fly ash at different reaction conditions.

It can be observed from Fig 5.1, that the oil conversion was not greatly affected by an increase in the reaction time. This implies that the reaction equilibrium had been reached after 5hrs when the reaction temperature was 160°C as the experiment was conducted under reaction conditions of 160°C, 15% catalyst loading and a methanol: oil ratio of 15:1.

Fig 5.2 shows that the highest conversion of 87.45% was obtained at a catalyst loading of 20% wt/wt which was comparative with the conversion value of 86.56% at 15%wt/wt. As a consequence, the optimal amount of catalyst used in this pilot was 15wt % loading as it was sufficient in catalyzing the reaction despite the challenge posed by the three phase reaction system where initial mass transfer control is inhibited due to heterogeneous kinetics.

The results shown in Figure 5.3 and from the activity profile, it was observed that as reaction temperature increased, there was a corresponding increase in the conversion of the triglycerides to methyl esters, showing peaks at temperatures ≥ 160°C. The highest conversion was 89.34% at a reaction temperature of 200°C which compares to the conversion value of 87.12% at the reaction temperature of 160°C. Influence of reaction temperature on oil conversion under reaction time of 5hrs, a catalyst loading of 15%wt/wt and a methanol: oil ratio of 15:1.

From the Figure5.4, it can be seen that the highest methyl ester conversion of 86.1% was observed at a methanol: oil molar ratio of 15:1. It can be also be observed from the activity profile that as the molar ratio increased from 6:1 to 15:1, the conversion was found to increase from 56.23 to 86.1%. Further addition of methanol into the reaction phase did not show any significant increase in the conversion. Moreover, a limitation to the use of a greater amount of methanol needed to propel the forward reaction is that an occurrence of flooding of the active sites by the methanol molecules rather than triglyceride molecules which might hinder the completion of the triglyceride conversion to methyl esters as conversion at methanol oil ratios higher than 15 appeared to decline [6]

Conclusion

The use of fly ash loaded with KNO_3 has been found suitable for the conversion of sunflower oil to methyl esters and glycerol. The fly ash based catalyst loaded with 5% wt KNO_3 at a reaction temperature of 160°C exhibits maximum FAME conversion (86.13%). The highest conversion of oil was obtained at a catalyst loading of 20%wt/wt. Influence of process parameters such as reaction time, reaction temperature and molar ratio of methanol to oil were also carried out. The oil conversion was not greatly affected by an increase in the reaction time. However, reaction time between 5 to 8 hours will lead to better oil conversion. The highest conversion was observed at a temperature of 200°C. The activity profile shows that increase in the methanol to oil ratio from 6:1 to 15:1 resulted in increased conversion from 52.23% to 86.1%. A decline in the oil conversion was observed when the recovered catalyst was recycled for use. The FAME produced was tested and found to meet ASTM standard.

REFERENCES

[1] Eskom fact sheet (2009). Ash management in Eskom; Generation fact sheets Available online at http://www.eskom.co.za/live/content.php?Item ID=28. Accessed May 2010.

[2] Maa F, Hannab MA. Biodiesel production: a review1. Bioresource Technology 1999:70:1-15.

[3] Schuchardta U, Serchelia R, Matheus R. Transesterification of Vegetable Oils: a Review. J. Braz. Chem. Soc 1998; 9(1):199-210.

[4] Georgogianni KG, Katsoulidis AP, Pomonis PJ, Kontominas MG. Transesterification of soybean frying oil to biodiesel using heterogeneous catalysts. Fuel Processing Technology 2009; 90:671–76

[5] Gitari, M., Petrik Leslie, Key D., Etchebers O. and Okujeni C. Mineralogy and Trace Element Partitioning in Coal Fly Ash/Acid Mine Drainage Co-Disposed Solid Residues.http://whocares.caer.uky.edu/wasp/AshSymposium/AshLibraryAgenda.asp (2005). (Accessed on the 24th March, 2010).

[6] Shu Q, Zhang Q, Xu G. Wang J, Preparation of biodiesel using s-MWCNT catalysts and the coupling of reaction and separation. Food and Bioproducts Processing 2009; 87:164-70.

*To whom all correspondence should be addressed.

Email address: obabajide@uwc.ac.za (O. Babajide)
Tel./Fax: +27 21 959 3878.

(182a) Electricity-Carbohydrate-Hydrogen (ECHo) Cycle for Sustainability

Percival Zhang and Weidong Huang, Virginia Polytechnic Institute and State University, Blacksburg, VA

We wish to suggest the electricity-carbohydrate-hydrogen (ECHo) cycle for sustainable development. In this energy conversion cycle, electricity is a universal energy carrier; hydrogen is a clean electricity carrier; and carbohydrate is a high-energy density hydrogen carrier (14.8 H_2 mass%) or a high-electricity storage carrier (~10 MJ electricity output/kg sugar) or a source for food, feed, liquid biofuels, and renewable materials. Each element of this cycle can be converted to the other reversibly & efficiently depending on resource availability, needs, and costs. To fill the gaps in this cycle, carbon dioxide can be biologically fixed by electricity or hydrogen for producing carbohydrate. These artificial photosynthesis processes are designed according to the design principles of synthetic biology, knowledge in the literature, and thermodynamics analysis; however, the validation experiments and its innovation require collaborative efforts from biologists, chemists, electrochemists, and engineers. Implementation of artificial photosynthesis on a large scale would solve numerous sustainability challenges, such as food production, CO_2 utilization, electricity storage, transportation fuel production, fresh water conservation, wildlife habitat restoration, or maintaining a small closed ecosystem for space travel and the survival of humankind.

See more of this Session: Unconventional Technologies for CO2 Capture, Conversion and Utilization II
See more of this Group/Topical: Topical G: Innovations of Green Process Engineering for Sustainable Energy and Environment

(182b) Novel Approach to CO2 Utilization for Power Generation Using An Electrochemical Reactor

Jose A. Vega, Casey Chartier, Shawna Smith and William Mustain, Chemical, Materials and Biomolecular Engineering, University of Connecticut, Storrs, CT

Recent interest in anion exchange membrane fuel cells (AEMFC) has been fueled mainly by the ability to use non-noble metal catalyst in alkaline media. Also, fuel oxidation does not require water and alcohol versions can operate on pure fuel, and anion electroosmotic drag suppresses fuel crossover. Finally, water management is potentially simplified since water is produced in the anode and partially consumed at the cathode. However, one of the main drawbacks of the hydroxide exchange membrane fuel cell is electrolyte degradation via Hofmann elimination and nucleophilic attack catalyzed by the presence of hydroxide. Operation in the carbonate cycle has been proven and suggested as a way of improving durability and power density [1,2]. In this work, the performance of an AEMFC was improved with the presence of CO2 in the cathode stream, suggesting that carbonate was being produced at the cathode [3]. This study found that use of carbonate ions does not affect chemical integrity and ionic conductivity of commercially available anion exchange membranes, while considerable degradation was found when using hydroxide anions. Hydrogen oxidation was possible using carbonate anions, following a 2-electron pathway and not affecting the exchange current density compared to hydroxide. Therefore, operation of a room temperature carbonate fuel cell that reacts oxygen and carbon dioxide to produce carbonate appears to be a viable option. The reactor will result in increased power output and electrolyte life compared to hydroxide exchange membrane fuel cells.

References: 1. L. Adams, S. Poynton, C. Tamain, R. Slade and J. Varcoe, ChemSusChem, 1 (2008) 79. 2. C. Lang, K. Kim and P. Kohl, Electrochem. Solid State, 9 (2006) A545. 3. J. Vega and W. Mustain, Elecrochim. Acta 55 (2010) 1638.

See more of this Session: Unconventional Technologies for CO2 Capture, Conversion and Utilization II
See more of this Group/Topical: Topical G: Innovations of Green Process Engineering for Sustainable Energy and Environment

A New Aqueous Mineral Carbonation Process Utilizing Bipolar Membrane Electrodialysis

Yoshinobu ABE[1], Hiroki NAGASAWA[1], Atsushi IIZUKA[2], Akihiro YAMASAKI[3] and Yukio YANAGISAWA[1]

[1] Department of Environment Systems, Graduate School of Frontier Sciences, the University of Tokyo,
5-1-5 Kashiwanoha Kashiwa-shi, Chiba 277-8563, Japan

[2] Research Center for Sustainable Science and Engineering, Institute of Multidisciplinary Research for Advanced Materials, Tohoku University,
2-1-1, Katahira, Sendai, Miyagi, 980-8577, Japan

[3] Department of Materials and Life Science, Faculty of Science and Technology, Seikei University,
3-3-1 Kichijoji-kitamach, Musashino-shi, Tokyo 180-8633, Japan

Keywords: mineral carbonation, bipolar membrane, electrodialysis, carbon dioxide

Abstract

A new aqueous mineral carbonation process which utilizes bipolar membrane electrodialysis was proposed. The proposed process consists of four unit operations, (a) acid leaching of alkaline metals (Ca^{2+} /Mg^{2+}) from mineral compounds by HNO_3, (b) Na_2CO_3 generation via absorption of CO_2 into NaOH, (c) Ca and Mg carbonate precipitation by addition of Na_2CO_3, and (d) HNO_3/NaOH regeneration from $NaNO_3$ solution with bipolar membrane electrodialysis. Regenerated HNO_3 solution is then recycled to the acid leaching step. In the Na_2CO_3 generation step, regenerated NaOH solution is contacted with the gaseous CO_2 to form Na_2CO_3 solution. Exhaust gas from power plant can be directly used without compression. Experimental studies were conducted to confirm the technical feasibility of this process. Results demonstrate that HNO_3 and NaOH were successfully regenerated from $NaNO_3$ solution by using bipolar membrane electrodialysis. This result suggests that we can utilize alkaline metals in mineral compounds by using acid and HNO_3 and NaOH solution can be used cyclically in this process.

1. Introduction

Carbon dioxide capture and storage (CCS) is a global warming mitigation technology that keeps carbon dioxide (CO_2) away from atmosphere. Typical CO_2 storage methods for CCS technologies are generally classified into three groups: geological storage, ocean storage, and mineral carbonation[1]. In the mineral carbonation method, carbon dioxide is sequestered in the form of stable calcium carbonate ($CaCO_3$) or magnesium carbonate ($MgCO_3$) by carbonation reaction of alkali-earth metal oxides contained in natural rocks. However, there are some problems in the existing techniques. The largest problem is a slowness of its carbonation reaction. Thus, several chemical and physical acceleration methods have been examined[2]. When we utilize acid leaching of alkaline metals from mineral compounds to accelerate this reaction, huge amount of acid are required to fix vast amount of

CO_2 and large amount of waste acid is produced. So we have to bear the cost of providing large amount of nitric acid and treatment of waste solution and these are very high. For these reasons, acid regenerate and waste solution treatement processes are needed. In order to overcome these problems, we proposed a new aqueous mineral carbonation process which utilizes strong acid and bipolar membrane electrodialysis. We conducted experimental studies to confirm the technical feasibility of this process.

2. Process description

We proposed a new aqueous mineral carbonation process. The proposed process consists of four unit operations (Fig.1). (a) acid leaching of alkaline metals (Ca^{2+}, Mg^{2+}) from mineral compounds and $Ca(NO_3)_2$, $Mg(NO_3)_2$ solution are produced, (b) Na_2CO_3 generation via absorption of CO_2 into NaOH, (c) carbonate precipitation by mixing of $Ca(NO_3)_2$, $Mg(NO_3)_2$ solution and Na_2CO_3, and (d) acid and alkaline regeneration from salt solution with bipolar membrane electrodialysis.

In the acid leaching step, alkaline metals are extracted from mineral compounds such as waste concrete, steel making slag and alkaline rocks. In this operation, HNO_3 solution is adopted as a leaching solution. In the Na_2CO_3 generation step, NaOH solution is contacted with the gaseous CO_2 to form Na_2CO_3 solution. In the carbonation sedimentation step, by adding Na_2CO_3 solution into the leaching solution, sediment of carbonate such as $CaCO_3$ or $MgCO_3$ is obtained and $NaNO_3$ solution is remained. In the acid and alkaline regeneration step, HNO_3 solution and NaOH solution are regenerated from $NaNO_3$ solution via bipolar membrane electrodialysis treatment. Regenerated HNO_3 and NaOH solution are then recycled in this process.

We use bipolar membrane electrodialysis to regenerate HNO_3 solution and NaOH solution. Bipolar membrane is composite membrane of cation and anion exchange membrane[3]. When voltage is applied, water can be split into proton and hydroxide ion by bipolar membrane. Sodium ions are transported through cation exchange membrane and react with hydroxide ion. While, nitrate ions are transported through anion exchange membrane and react with Protons. So we can regenerate HNO_3 solution and NaOH solution.

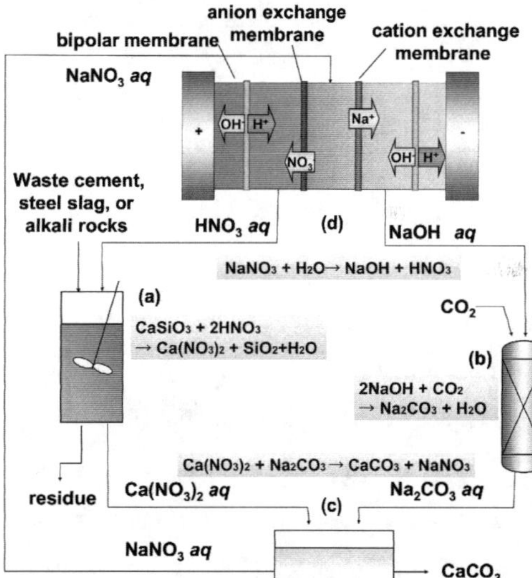

Fig.1 Outline of this process

3. Materials and Methods

3-1 Characterization of samples

Waste concrete (Calcium silicate hydrate gel, $Ca(OH)_2$, SiO_2, $3CaO{\cdot}2SiO_2{\cdot}4H_2O$), Blast furnace slag, Wollastonite ($CaSiO_3$) and Olivine (Mg_2SiO_4) samples were used as a source of alkali-earth metal oxides in the study. The elemental composition of the waste concrete and alkali rocks were analyzed with an X-ray fluorescence apparatus (XRF, Rigaku, ZSX Primus II). Particle size distribution of the samples was measured with a laser-scattering particle size measurement apparatus (Shimadzu, SALD2100). **Table 1** summarizes the elemental compositions and particle size distributions of alkali wastes and alkali rocks samples. When one metric ton of the waste concrete, blast furnace slag, wollastonite and olivine are completely carbonated, 0.26, 0.29, 0.32, 0.47 metric ton of CO_2 can be sequestered, respectively.

Table1 Elmental compositions and size distribution the alkali wastes and rocks

Alkali rocks and waste	Elemental composition(wt%)						Size distribution
	Ca	Mg	Si	Fe	Al	other	of sample(μm)*
Wollastonite	29.2	0.1	13.5	0.6	0.1	56.6	5-100(66.7)
Olivine	0.9	25.7	24.7	11.5	0.7	36.3	5-100(66.7)
Waste cement	23.6	0.5	9.2	3.1	2.4	63.2	10-230(188.9)
Blast furnace slag	26.3	1.9	8.8	0.4	4.0	62.4	0.2-100(0.37)

*volume-based light scattering diameter, modal diameter are in parentheses

3-2 Calcium and Magnesium ion extraction experiments

Calcium and magnesium ion extraction experiments were conducted. In this study, influences of concentration of acid (0.1, 0.5, 1.0, 2.0 M) and under stoichiometric conditions (amount of Ca in waste concrete and nitric acid) were studied. Fundamental condition was set as follows: initial weight ratio of waste concrete and alkali rocks to water (1 g / 250 mL) and temperature (25 °C). In under stoichiometric condition study, initial weight ratio of waste concrete is 1.666 g / 250 mL and concentration of acid is 0.1 M. During the extraction experiments, the extracted solutions were sampled at given intervals. Then, the concentration of calcium, magnesium, silicon, iron and aluminum ions of the extracted solution were determined by inductively coupled plasma atomic emission spectrometry ICP-AES (HITACHI, P-4010).

3-3 Carbonate precipitation experiments

Carbonate precipitation experiments were conducted. Blast furnace slag sample was used as a source of calcium in the study. Calcium ion extraction was conducted under calcium excess condition (initial weight ratio of blast furnace slag to acid is 30 g / 1 M HNO_3 200 mL). After calcium extraction, $Ca(NO_3)_2$ solution and residue are separated by suction filtration. By adding 0.5 M Na_2CO_3 solution into the $Ca(NO_3)_2$ solution, sediment of $CaCO_3$ is obtained. The precipitated $CaCO_3$ was analyzed by X-ray diffraction (XRD, Rigaku, MiniFlex II).

3-4 Acid and alkaline regeneration experiments

In the acid and alkaline regeneration step, HNO_3 solution and NaOH solution are regenerated from $NaNO_3$ solution via bipolar membrane electrodialysis treatment. We used $NaNO_3$ solution which remained in the carbonation sedimentation step (0.5 M $NaNO_3$) and separated $CaCO_3$ and $NaNO_3$ solution by suction filtration. A laboratory-scale electrodialysis apparatus was applied in this study. This apparatus contains three cell pairs (feed, acid, base cell) between two electrodes at the both end. Then, current, voltage, pH value were analyzed.

4. Results and Discussion

4-1 Calcium and Magnesium ion extraction experiments

4-1-1 Influence of the concentration of acid

Figure 2 and 3 shows the influence of the concentration of nitric acid.

【waste concrete】

When the concentration of nitric acid was 0.1 M, calcium ion extraction ratio was increased about 80% after 10 min extraction. The concentration of silicon, iron and aluminum ions extraction ratio were about 50, 75, 35 %, respectively. These extraction ratio were similar extent under other experimental conditions.

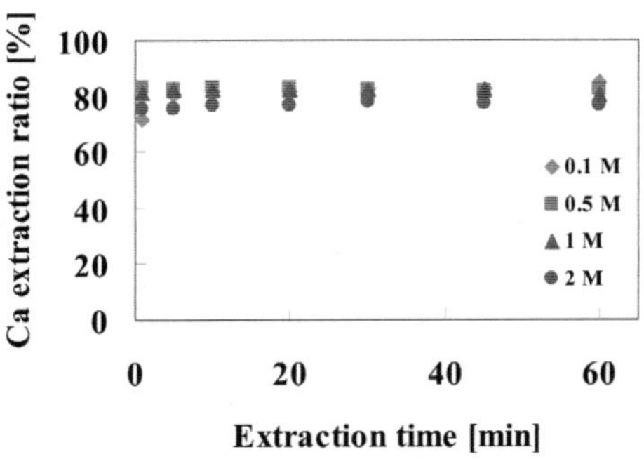

Fig.2 Time variation of extraction ratio of Ca from waste concrete for various concentration of nitric acid.

【alkali rocks】

For wollastnite, when the concentration of nitric acid was 0.1 M, calcium ion extraction ratio was about 12% after 60 min extraction. When the concentration of nitric acid was 0.5 to 2 M, calcium ion extraction ratio was increased about 100% after 60 min extraction. For olivine, although magnesium ion concentration increased with an increase of the concentration of nitric acid, their concentrations were low compared with that of wollastonite.

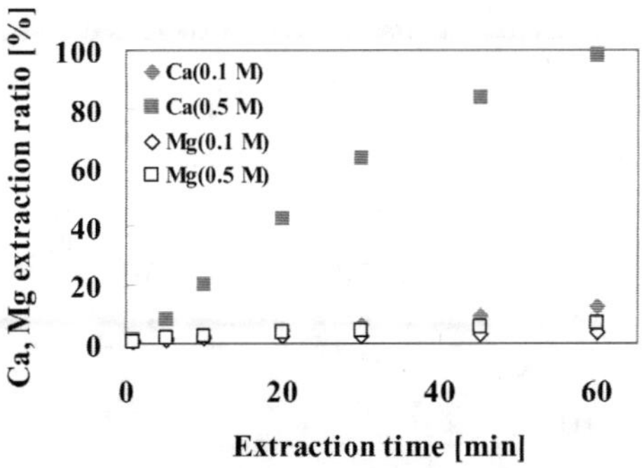

Fig.3 Time variation of extraction ratio Ca from wollastonite and Mg from olivine.

3-1-2 Under stoichiometric conditions

Figure 4 shows the result of the calcium ion extraction experiment under stoichiometric conditions. When the concentration of nitric acid was 0.1 M, calcium ion extraction ratio was increased about 80% after 10 min extraction. The concentration of silicon ions extraction ratio was about 40 %. We can confirm that calcium ion extraction ratio was similar extent under various concentration of nitric acid conditions and other ions extraction ratio were decreased.

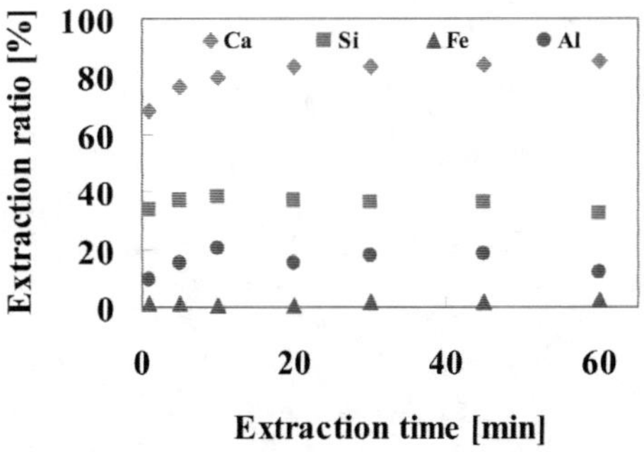

Fig.4 Time variation of extraction ratio of Ca, Si, Fe and Al from waste concrete under stoichiometric conditions.

4-2 Carbonate precipitation experiments

Figure 5 shows XRD pattern of the precipitated carbonate. This result indicates that sediment of $CaCO_3$ was produced by adding Na_2CO_3 solution into the $Ca(NO_3)_2$ solution.

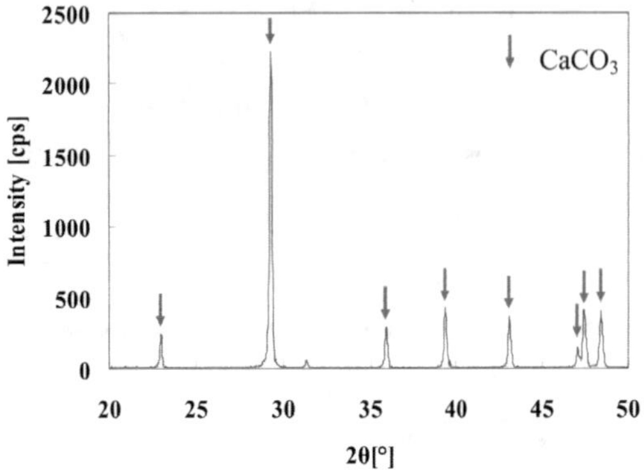

Fig.5 XRD chart of calcium carbonate

4-3 Acid and alkaline regeneration experiments

【Time variation of current and voltage】

Figure 6 shows Change in current and voltage value. Current value increased and reached 2.0 A after 2 min. Meanwhile, voltage value decreased and reached about 11.5 V after 4 min. After 30 min, current value dramatically decreased and voltage value increased. Since the concentration of $NaNO_3$ solution reached about 0 M, these results were caused.

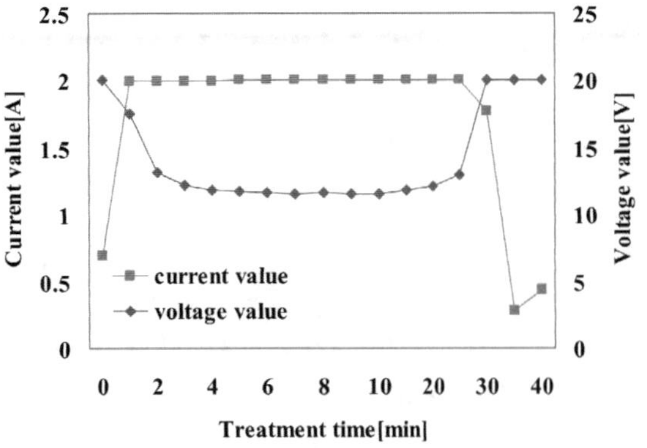

Fig.6 Time variation of current and voltage

【Time variation of pH】

Figure 7 shows time variation of pH. In acid cell, pH value dramatically decreased and reached 1.68 after 2 min. In base cell, pH value was moderately increased and reached 13.51 after 4 min. Time variation of pH in acid cell and base cell indicated that HNO_3 and NaOH were successfully regenerated from $NaNO_3$ solution.

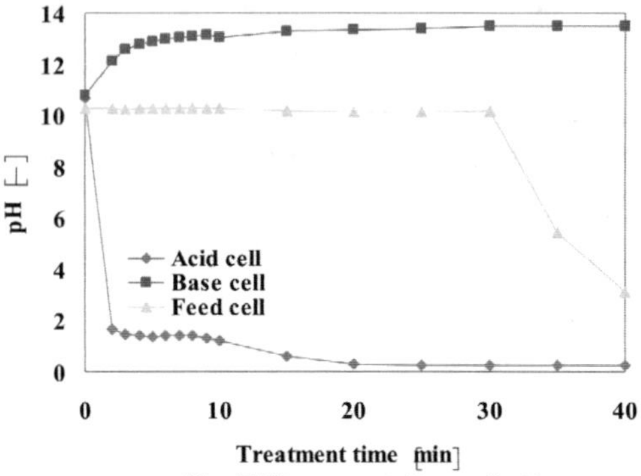

Fig.7 Time variation of pH

5. Conclusions

We proposed a new aqueous mineral carbonation process which utilizes strong acid and bipolar membrane electrodialysis. Experimental studies were conducted to confirm the technical feasibility of this process. Results demonstrate that HNO_3 and NaOH were successfully regenerated from $NaNO_3$ solution by using bipolar membrane electrodialysis. This result suggests that acid and alkaline solution can be used cyclically in this process. These results indicates the technical feasibility of this process.

Reference

[1] IPCC, IPCC special report on carbon dioxide capture and storage, *Cambridge University Press*, 2005.
[2] J. Sipila *et al.*, *Report VT 2008-1*, 2008.[3] R. Simons, *Journal of Membrane Science*, 78, 13-23, 1993

(182d) Functionalized Polyaniline COMPOSITE PARTICLES for CO_2 Sequestration and FERTILIZER Conversion

Yong Min, Nanomaterial Innovation Ltd., Columbus, OH and J. James Lee, The Ohio State University, Columbus, OH

Abstract This work is to develop low-cost, thermally stable and high CO_2 affinity conducting polymers and their association compounds (i.e., dopants) as a "molecular CO_2 reservoir" to not only achieve effective CO_2 capture under harsh conditions as in an electric power plant but also convert the sequestered CO_2 into useful products such as NH_4HCO_3 fertilizer in a simple reaction and allow the "molecular CO_2 reservoir" to be re-used in a continuous process.

Novel functionalized polyaniline was synthesized through in-situ polymerization with various dopants and particles. The material composition and CO_2 sequestration were investigated through various analytical techniques, such as, X-Ray diffraction (XRD), SEM, TEM, dielectric measurement, UV/vis, Rubotherm high pressure/vacuum absorption, XPS, elemental analysis, and FTIR.

Those composite particles have superior capabilities to sequestrate not only CO_2 but also SO_x and NO_x at low temperatures. Studies were also conducted to demonstrate the conversion of the sequestrated CO_2 into fertilizers such as NH_4HCO_3, NH_4HSO_4, and NH_4NO_3 at ambient conditions. The reaction yields, product purity, thermal dynamic and kinetic of the reaction were also investigated. The performance of our materials is better than the best exiting amine-based CO_2 sorbent.

The new polyaniline composite particles and associated processing techniques may lead to a low-cost CO_2 sequestration process in the industry that can permanently remove CO_2 from the emitting sources.

Keywords: Polyaniline, CO_2 sequestration, fertilizer conversion.

See more of this Session: Unconventional Technologies for CO_2 Capture, Conversion and Utilization II
See more of this Group/Topical: Topical G: Innovations of Green Process Engineering for Sustainable Energy and Environment

(182e) Amine Promotion of CO_2 Conversion for Artificial Photosynthesis

Wei Zhu, Chemical and Biomolecular Engineering, University of Illinois, Urbana-Champaign, Urbana, IL, Brian A. Rosen, Department of Chemical and Biomolecular Engineering, University of Illinois at Urbana-Champaign, Urbana, IL and Richard I. Masel, Chemical and Biomolecular Engineering, University of Illinois at Urbana-Champaign, Urbana, IL

The objective of our work is to learn how to use artificial photosynthesis to create feedstocks for renewable fuels. We imagine collecting electricity via a series of mirrors and a solar panel and using the electricity to convert CO2 and water to CO, H2 and formic acid. At present the limiting factor in these processes is the ability to convert CO2. So far one can only obtain reasonable efficiencies by using a catalyst like mercury at high overportentials, and even then the efficiency is low. Instead most of the electrons go to electrolyzing water.

We have been exploring the use of amines to suppress water electrolysis and enhance CO2 electrolysis. We find that choline chloride raises the onset of hydrogen evolution by about 0.5 volt, and lowers the overpotential for CO2 conversion by a similar amount. The result is that there is a region of potential where CO2 electrolysis is more rapid than hydrogen evolution. The products vary with the catalyst with CO evolution dominating on nickel catalysts, and formic acid dominating on palladium catalysts. More work needs to be done, but it appears that this work will lead to a practical route for artificial photosynthesis.

See more of this Session: Unconventional Technologies for CO2 Capture, Conversion and Utilization II
See more of this Group/Topical: Topical G: Innovations of Green Process Engineering for Sustainable Energy and Environment

(182f) Electrochemical Conversion of CO_2 to Syngas in An Electrochemical Reactor

Devin T. Whipple and Paul J.A. Kenis, Chemical and Biomolecular Engineering, University of Illinois at Urbana-Champaign, Urbana, IL

Significant reductions in carbon dioxide emissions and the development of non-fossil fuel alternative energy sources is critical to both minimize the effects of CO2 as a green-house gas and reduce our dependence on imported non-renewable energy. In order for renewable and carbon neutral energy sources to replace fossil fuels, technological advances are needed that will allow their penetration into the energy market. The electrochemical reduction of CO2 has to potential to help overcome several of the challenges facing the implementation of carbon neutral energy sources. The electrochemical conversion of CO2 into fuels allows the storage of electrical energy in chemical form, which is important considering that currently over 75% of carbon neutral fuels produce electricity. Storing this electrical energy is high density chemical form would facilitate the penetration of carbon neutral energy into the transportation sector, which makes up a large portion of our energy use but has only 3% supplied by renewable sources. Furthermore, the ability store electricity would aid the expansion of popular renewable sources such as wind and solar that are intermittent and require output leveling. In particular, the production of CO from CO2 is attractive due to its versatility as a feedstock in Fischer-Tropsch synthesis to a variety of products including liquid hydrocarbon fuels. However, before the electrochemical reduction of CO2 can be viable, significant improvements in the energy efficiency and current densities of the process need to be achieved. One of the main obstacles to this is high overpotentials due the formation of the CO2.- intermediate. Significant improvements could be made by finding conditions (such as electrolytes) and catalysts that stabilize this intermediate. We have developed a microfluidic reactor that employs a flowing electrolyte in lieu of a membrane and allows greater flexibility in reactor conditions. In this presentation we will report our studies investigating the effects of electrolyte pH, electrolyte species, and temperature on the energy efficiency and current density of the electrochemical reduction of CO2 to CO.

See more of this Session: Unconventional Technologies for CO2 Capture, Conversion and Utilization II
See more of this Group/Topical: Topical G: Innovations of Green Process Engineering for Sustainable Energy and Environment

(182g) Enhanced Hydrogen Production with Integrated Carbon Sequestration Using Mg(OH)2 Sorbent

Kyle J. Fricker, Earth and Environmental Engineering, Columbia University, New York, NY and Ah-Hyung Alissa Park, Earth and Environmental Engineering & Chemical Engineering, Columbia University, New York, NY

Hydrogen production is vital to our energy future, particularly as a raw material in synthetic fuel synthesis. Synthesis gas, generated from the gasification of carbonaceous material, typically has a CO: H2 ratio of approximately 2:1 while a ratio of 1:2 is desired for liquid fuel synthesis. In order to achieve the desired CO:H2 ratio, large catalytic reactors are often used to produce additional H2 via the industrial water-gas-shift (WGS) reaction. Some recent WGS system reaction designs integrate carbon capture technology. The removal of CO2 generated during WGS reaction, through CO2 capture (e.g., membrane or high temperature sorbent technologies), will shift the equilibrium towards the products; thus, enhancing the hydrogen yield. Ca-based sorbents seem to achieve the enhanced hydrogen production successfully, yet these materials are not a permanent carbon storage methods since they likely originate from naturally occurring carbonates like limestone. Thus, CO2 capture using Ca-based sorbents must regenerate the sorbent, adding the energy penalty and associated cost. In this study, we focus on another metal oxide sorbent, magnesium hydroxide (Mg(OH)2) because unlike Ca-based sorbent Mg(OH)2 does not require sorbent regeneration. Gas-solid carbonation of a Mg(OH)2 sorbent is thermodynamically favored within the temperature range associated with WGS reaction. Theoretically, when coupled with the WGS, the hydrogen conversion can sustain near 100% until the calcination temperature of MgCO3 is reached; thereby eliminating the need for the larger and inherently more costly, Low-Temperature Shift (LTS) reactor. Previous research into the carbonation of Mg(OH)2 showed that kinetics limit this reaction, and the exact reaction pathway is not well understood. Some have suggested that the dehydroxylation-carbonation process is very much path dependent. Thus, this study focuses on the synthesis of an engineered Mg(OH)2 sorbent with controlled mesoporous structures to provide maximum conversion during gas-solid reactions, while identifying the reaction rate and mechanisms of various reaction paths. The ultimate goal is to utilize abundant magnesium-silicate minerals like serpentine and olivine to synthesize a Mg(OH)2 sorbent that is both environmentally benign and highly reactive. This sorbent has the ability to fundamentally improve the industrial WGS reaction process while also sequestering CO2 that would otherwise have been vented to the atmosphere.

See more of this Session: Unconventional Technologies for CO2 Capture, Conversion and Utilization II
See more of this Group/Topical: Topical G: Innovations of Green Process Engineering for Sustainable Energy and Environment

Multi-Scale Methodology for Process Engineering Innovations

Jinghai Li* and Ge We**
State Key Laboratory of Multi-Phase Complex Systems,
Institute of Process Engineering,
Chinese Academy of Sciences,
Beijing 100190, China
*email: jhli@home.ipe.ac.cn, **presenting author, email:wge@home.ipe.ac.cn

Challenges in quantitative design of chemical processes mainly reside in their complex structures, which are usually multi-scale in nature, and are difficult to analyze by the average approach as commonly used. The multi-scale objective variational (MOV) methodology is promising in coping with complex systems and sheds light on innovations of new and/or improved chemical processing technologies for sustainable energy and environment. This presentation will provide an overview of four critical issues in the MOV methodology, (i) correlation between phenomena at different scales, (ii) compromise between different dominant mechanisms, (iii) coupling between spatial and temporal structural changes, (iv) critical phenomena occurring in complex systems. Impacts of this methodology on process engineering innovation will be elaborated with nearly three-decade development at Institute of Processing Engineering (IPE) from a rough idea to modeling approaches, software, supercomputing hardware and finally to industrial applications as well as its extension to a number of chemical and physical systems. Possibility of realizing virtual process engineering based on the accuracy and efficiency of this methodology and dramatic changes in high-performance computing will be prospected finally.

Key words: HPC, multi-phase flow, multi-scale, simulation, virtual process engineering

(228b) Artificial Photosynthesis: The Future of Renewable Fuels and Chemicals

Rich Masel[1], Brian A. Rosen[2], Wei Zhu[3] and Amin Salehi-Khojin[2], (1)University of Illinois, Urbana-Champaign, Urbana, IL, (2)Department of Chemical and Biomolecular Engineering, University of Illinois at Urbana-Champaign, Urbana, IL, (3)Chemical and Biomolecular Engineering, University of Illinois, Urbana-Champaign, Urbana, IL

This presentation will provide an overview and research vision of catalytic and reaction engineering technologies for conversion of CO2 into re-usable energy and/or products at productivity and energy efficiency substantially higher than the natural, photosynthesis process.

See more of this Session: International Forum On Clean and Renewable Energies (plenary)
See more of this Group/Topical: Topical G: Innovations of Green Process Engineering for Sustainable Energy and Environment

(228c) Electrochemical Energy Storage and Conversion System Integration and Application for New Energy Vehicle and CO2 Utilization

Zi-Feng Ma, Jiani Shen, Jin-Kun Li and Yinmei Yin, Department of Chemical Engineering, Shanghain Jiao Tong University, Shanghai 200240, China

Energy storage systems are very important for the new energy vehicle. Electric vehicles (EVs) including the fuel cell electric vehicle (FCEV) are one of the most important new energy vehicles. The obstacle to the commercialization for EVs is hydrogen storage and electricity storage with low cost, high density and safety. According to the expertise from the demonstration of the FCEV and PHEV in the world, all-electricity drive motor system which integrated with advanced battery or supercapacitor with fuel cell stack will become the promising power system for the solution of green and sustainable vehicle. Fuel cells will play an important role in the new energy vehicle. The National Basic Research Program (973 Program) is China's on-going national keystone basic research program, which was approved by the Chinese government in June 1997 and is organized and implemented by the Ministry of Science and Technology (MOST) of China. Based on the demonstration activities of the FCEV and other new energy vehicles during 2008 Beijing Olympic and 2010 Shanghai EXPO, we found that it is important to study the lower cost materials and the process engineering issue for fuel cell electric vehicle. At the meantime, one 973 project titled of "Basic study to energy storage system (Hydrogen & electricity storage) with low cost and high power density for electric vehicle" was started since 2007. Many research groups those come from different universities and institute were involved in this project. In this project, the researches focused on several topics, such as molecular design and synthesis of hydrogen storage materials with high capacities, fundamental studies of lower cost electrochemical supercapacitors and advanced secondary batteries and basic study of solid-state electrolytes and related electrode materials for solid oxide fuel cell. The applied fundamental research on electrode materials and function in CO2 emission reduction and utilization of solid oxide cell were also involved. This talk will present some of the research progress of these projects in detail. Acknowledgement: The authors are grateful for financial support of this work by the 973 Program of China (2007CB209700), the STCSM of China (09XD1402400, 10520708900).

See more of this Session: International Forum On Clean and Renewable Energies (plenary)
See more of this Group/Topical: Topical G: Innovations of Green Process Engineering for Sustainable Energy and Environment

(228d) Recent Research Development In Biomass Conversion to Advanced Bio-Fuels

Johnathan E. Holladay, Manager, Biomass Business Sector, Pacific Northwest National Lab, Richland, WA

The United States is abundant in biomass resources. The Pacific Northwest National Laboratory (PNNL) have been conducting research on conversion of biomass into chemical products and advanced fuels for several decades. In this presentation, the author will discuss significant research advancement, current status, and challenges. PNNL's participation and research directions of two recent DOE-funded consortia, National Advanced Biofuels Consortium, the National Alliance for Advanced Biofuels and Byproducts - an Algal Biofuels Consortium, will be shared.

See more of this Session: International Forum On Clean and Renewable Energies (plenary)
See more of this Group/Topical: Topical G: Innovations of Green Process Engineering for Sustainable Energy and Environment

(295a) Ionic Liquids: From Fundamental Researches to Industrial Applications

Suojiang Zhang, Research Laboratory of Ionic Liquids and Clean Energy, Institute of Process Engineering, Chinese Academy of Sciences, Beijing, China and Xiaomin Liu, Institute of Process Engineering, Chinese Academy of Sciences, Beijing, 100080, P. R. China, Beijing, China

Ionic liquids (ILs) are a new class of substances exclusively composed of cations and anions, and they have shown unique physico-chemical properties differing from aqueous ionic solutions and high temperature molten salts. Although many attentions have been paid on ILs, the knowledge of ILs is still limited to experimental try-and-errors and experiences, and it is far away from our need for understanding the nature of ILs, especially for the industrial applications. Therefore, it is indispensably important to study the structure-property relationships and the scale-up principles in the ILs systems, which is the scientific frontier in chemistry and chemical engineering.[1] ILs database have been established, including 2078 kinds of ILs, 41 kinds of physical properties and 54,803 pieces of data. A "Roadmap" was built to screen suitable ILs from numerous combinations of 276 kinds of cations and 55 kinds of anions. A QSPR (quantitative structure property relationship) model was developed for predicting the properties of ILs. Hydrogen bond network structure and its three-dimensional expansibility in ILs were revealed. [2]Molecular force fields for series of ILs were developed, in terms of which molecular dynamic simulation was performed to investigate extensively the micro-structures, interactions and properties. [3] Besides, a set of novel equipment for in-situ measurement of the transport-reaction coupling performance in ILs was established based on the electric conductive and dynamic characteristics of ILs. [4] Based on the fundamental research on ILs, an innovative cleaner process of methyl methacrylate, the monomer of perspex, was developed to replace the traditional highly toxic hydrocyanic acid process. [5] In terms of "ionic switch" principle, a novel de-acidification process of lubricant base oil has been developed, which overcomes many disadvantages such as high energy consumption, heavy pollution in the traditional caustic washing process. Another example is using ILs as catalyst for producing ethylene glycol instead of the traditional direct hydration of ethylene oxide, the new process could save large quantity of energy and significantly reduce wastewater discharge amount. [6] Acknowledgment This work was supported by the National Science Fund of China for Distinguished Young Scholar (20625618), National Basic Research Program of China (973 Program, 2009CB219900), Knowledge Innovation Program of the Chinese Academy of Sciences (KGCX2-YW-321), National Key Technology R& D Program(2008BAF33B04)and National Natural Scientific Fund of China (50974113, 20903098). References [1] Zhang S, Lu X, Zhou Q, et al. "Ionic Liquids: Physicochemical Properties", Elsevier, 2009. [2] Dong K, Zhang S, Wang D, et al. J. Phys. Chem. A, 2006, 110(31): 9775. [3] Liu X, Zhang S, Zhou, G, et al. J. Phys. Chem. B, 2006, 110(24): 12062. [4] Dong H, Wang X, Liu L, et al. Chem. Eng. Sci., 2010, 65: 3240. [5] Y. Diao, R. Yan, S. Zhang, et al. J. Mol. Catal. A: Chem., 2009, 303: 35. [6] J. Sun, W. Cheng, W. Fan, et al. Catal. Today, 2009, 148: 361.

See more of this Session: Novel Catalytic and Separation Process Based On Ionic Liquids

(295b) Artificial Photosynthesis: Conversion of CO2 at Low Temperature and Potential Using Ionic Liquids

Brian A. Rosen, Department of Chemical and Biomolecular Engineering, University of Illinois at Urbana-Champaign, Urbana, IL, John Haan, Chemical Engineering, University of Illinois at Urbana-Champaign, Urbana, IL, Wei Zhu, Chemical and Biomolecular Engineering, University of Illinois, Urbana-Champaign, Urbana, IL and Rich Masel, University of Illinois, Urbana-Champaign, Urbana, IL

Alternative energy and climate change are among the largest concerns facing scientists and engineers at the start of the twenty-first century. The focus of this research is the design of a system that can efficiently capture atmospheric carbon dioxide and reduce it to CO, a precursor for fuel synthesis. Rather than traditional solvents such as water, sulfuric acid, or acetonitrile, room temperature ionic liquids (RTIL's) will be employed as the solvent for CO2 capture and reduction. RTIL's have many favorable properties such as extended electrochemical windows, high thermal stability, high electrical conductivity, and a 100+ fold increase in carbon dioxide solubility compared to water. Additionally, the absence water can increase the efficiency of CO production by eliminating the hydrogen and oxygen evolution reactions. Cyclic voltammetry has shown the catalytic reduction of CO2 at the Pt and Pt/Ru surface in imidazolium based ionic liquids such as EMIM BF4, BMIM BF4 and EMIM NTF. Cathodic shoulders correlating to the reduction of CO2 are found at ca. -0.5V vs. SHE, which is over half a volt less negative than previously reported in aqueous solvents, and 400mV less negative than reported in other non-aqueous solvents. CO stripping experiments and Broadband Sum Frequency Generation Spectroscopy (BB-SFG) has qualitatively confirmed that at least one of the reduction products is CO. Because this novel setup has eliminated much of the overpotential traditionally associated with CO2 reduction, this process can take place at much higher efficiencies, as well as re-open the door for investigating non-precious metal electrodes originally eliminated due to their low efficiency in the presence of water.

See more of this Session: Novel Catalytic and Separation Process Based On Ionic Liquids
See more of this Group/Topical: Topical G: Innovations of Green Process Engineering for Sustainable Energy and Environment

(295c) Gas Sorption in Ionic Liquid-Carbon Nanotube Composite Materials

Wei Shi, Computational Chemistry, National Energy Technology Laboratory, Parsons Corporation, Pittsburgh, PA and Dan C. Sorescu, Computational Chemistry, National Energy Technology Laboratory, Pittsburgh, PA

Molecular dynamics and Monte Carlo simulations were used to study the ionic liquid 1-n-Hexyl-3-methylimidazolium bis(Trifluoromethylsulfonyl)amide ([hmim][Tf2N]) confined into the (20,20) and (9,9) carbon nanotubes and gases (CO_2, H_2) absorption into the composite material consisting of the ionic liquid and the carbon nanotube. Structures of the ionic liquid confined in the carbon nanotubes will be presented. The self-diffusivity of the gases and the ionic liquid in the carbon nanotube have been calculated. For gas sorption, the highest amounts of adsorbed CO_2 and H_2 are observed in carbon nanotube, followed by the composite material, and the ionic liquid exhibits the least amounts of absorption. The CO_2/H_2 sorption selectivity exhibits the highest value in the composite material, followed by the ionic liquid, and the carbon nanotube exhibits the least selectivity. When the temperature is increased, the CO_2 sorption decreases in all three types of materials. In contrast, the H_2 sorption increases in [hmim][Tf2N], decreases in the carbon nanotube, and almost does not change in the composite material. Novel absorption behavior for CO_2/IL/carbon nanotube and differences from those of CO_2/IL will be shown.

See more of this Session: Novel Catalytic and Separation Process Based On Ionic Liquids
See more of this Group/Topical: Topical G: Innovations of Green Process Engineering for Sustainable Energy and Environment

Ionic Liquids Assisted Preparation of Cellulose Coated Bacterial Biosorbents

Yeoung-Sang Yun[1,2], Thuy T.P. Pham[1], Myung-Hee Song[1], Sung Wook Won[1], Robin D. Rogers[2]

[1]Environmental Biotechnology National Research Lab., School of Chemical Engineering and Department of BIN Fusion Technology, Chonbuk National University, Jeonju, Chonbuk 561-756, South Korea

[2]Center for Green Manufacturing and Department of Chemistry, The University of Alabama, Tuscaloosa, AL 35487, USA

Biosorbents prepared from natural sources have a great potential as low-cost competitive sorbents able to bind ionic species in solution phases. However, their mechanical strength and chemical stability especially under acidic condition should be improved to be used for the practical application. This study aimed at solving these bottlenecks with help of cellulose dissolved in ionic liquids. Waste bacterial biomass generated from an amino acid fermentation industry was used as raw material. The raw biomass was chemically modified and immobilized with chitosan for increasing the binding sites for ionic solutes. To make process applications possible, the chemically modified biosorbents were molded as spherical and fiber shapes. A simple but effective way to coat cellulose on the surface of biosorbents was developed and will be presented. Briefly, the half-dried biosorbents were brought into contact with dissolved cellulose in ionic liquids. Then water was left to release from the inside to the surface of biosorbents, and cellulose surrounding the biosorbents began to precipitate on the surface, making the thin cellulose film. Resulting cellulose-coated biosorbents had enough mechanical strength to be applied in column-type rectors. Furthermore, the developed biosorbents were stable in acidic and chloride-rich solutions. Successful applications of the developed biosorbents for removal and/or recovery of toxic or precious solutes from the aqueous phase will be demonstrated in this presentation.

(295e) High Absorption Performance of HCl in 1-Octyl-3-Methylimidazolium Chloride

Shuo Li, Zhen Zhang, Xiangping Zhang and **Suojiang Zhang**, Research Laboratory of Ionic Liquids and Clean Energy, Institute of Process Engineering, Chinese Academy of Sciences, Beijing, China

Incineration of wastes and combustion of chlorine-rich coal are the major sources of local and global environmental pollution related to hydrogen chloride (HCl) emission. At the same time, HCl is a common tail gas in chemical industry such as PVC and Freon production. HCl is regarded as the most significant substance to destroy the ozone layer, moreover it is harmful, corrosive and can contribute to the production of chlorinated unburned hydrocarbons and dioxins. In order to capture HCl for reducing the environmental risk and utilizing HCl as chloride resource, the investigation of efficient methods for HCl removal and recovery is critically important in recent decades.

Although there are some commercial processes for flue gas dechlorination, such as hydrated lime, ion-exchange resin, condensation, alkali absorption, water absorption etc., the disadvantages of these processes are obvious, such as low efficiency, absorbent waste, secondary pollution and so on. Therefore, a novel solvent that could facilitate the separation of HCl from flue gas without concurrent loss of the absorbent into the gas stream is highly required. In this regard, ionic liquids (ILs) show great potential as an alternative for such applications. In recent years, significant progress has been made in the application of ILs as alternative solvents due to their unique properties such as negligible vapor pressure, a broad range of liquid temperatures, excellent thermal and chemical stabilities, tunable physicochemical characteristics and selective absorption of certain organic and inorganic materials, such as CO_2, SO_2 , NH_3 and so on.

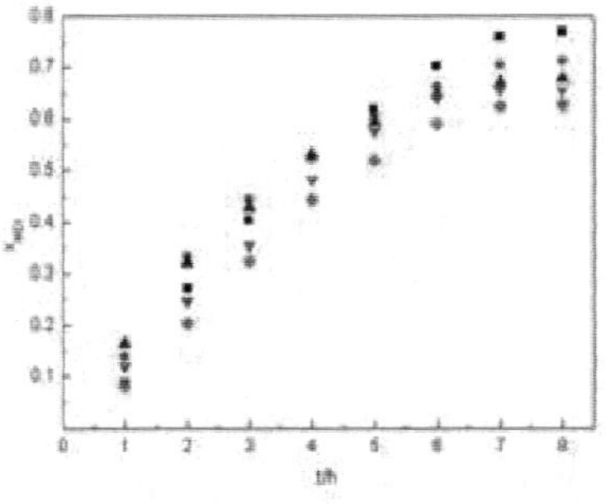

Figure 1. HCl Absorption in [Omim] Cl at 101 kPa: ■, 273 K; •, 293 K; ▲, 313 K; ▼, 333 K; ◆, 353 K.

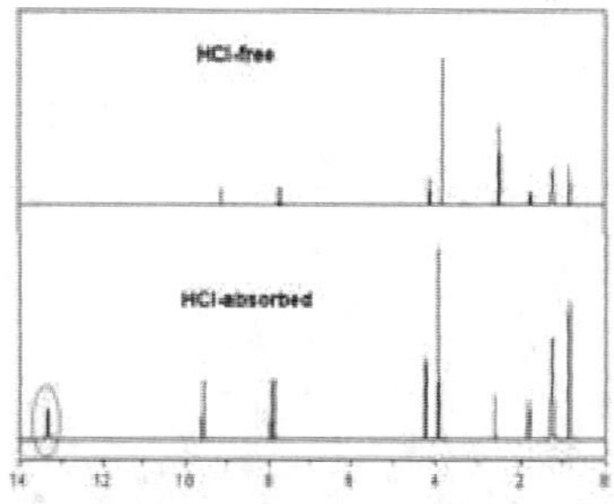

Figure 2. ^{1}H NMR spectra of HCl-free and HCl-absorbed [Omim] Cl

In order to develop a novel process for HCl absorption, it is necessary to have the knowledge of HCl absorption in ILs. In this work, 1-octyl-3-methylimidazolium chloride ([Omim]Cl) was synthesized, and the HCl absorption was examined at 273 K, 293 K, 313 K, 333 K, and 353 K under atmospheric pressure. The results are presented in Figure 1. It was found that the HCl absorption properties of these ILs at atmospheric pressure are quite high. For example, the mole fraction of HCl in water is only 0.2886 at 273 K, while in [Omim] Cl is 0.7697. And the saturated absorption of HCl decreases when the temperature increases. The recycle experiments were repeated for four times at 293 K and atmospheric pressure. It is found that the decreasing of the absorption ability was very little. The recovery rates in four times were 99.5 %, 99.1 %, 98.3 %, and 97.7 %, which indicate that ILs can be recycled and used repeatedly. ^{1}H NMR spectra of HCl-absorbed [Omim] Cl were compared with those of HCl-free samples. It is shown that in the ^{1}H NMR spectra of HCl-absorbed samples, there is a new absorption peak at about 13.32 ppm (specific absorption peak of $ClHCl^-$). The results are presented in Figure 2. As a result, the high absorption performance of HCl in [Omim] Cl could be ascribed to three factors: the formation of $ClHCl^-$; the hydrogen bonds between C-2 active hydrogen on the imidazol ring and HCl; and the high asymmetry between anions and cations.

Comparing with the conventional absorbents, the [Omim] Cl have several powerful advantages: higher ability of HCl absorption; recyclable and reusable; the absorbed HCl can be recovered and utilized. Therefore, it shows great potential as alternative solvents for HCl absorption in the chemical industries, and more effective ionic liquids for absorbing HCl are expected to be researched in future.

The author gratefully acknowledges the support of Ministry of Science and Technology of China (973 program 2009CB219901), the Natural Science Foundation (20803075).

See more of this Session: Novel Catalytic and Separation Process Based On Ionic Liquids
See more of this Group/Topical: Topical G: Innovations of Green Process Engineering for Sustainable Energy and Environment

(295f) Adsorption Separation of Reaction Product From Ionic Liquid Reaction Mixture for Conversion of Biomass Into 5-Hydroxymethylfurfural (HMF)

Wei Liu[1], Feng Zheng[2], Heather Brown[2] and Allan Cooper[2], (1)Energy and Environmental Directorate, Pacific Northwest National Lab, Richland, WA, (2)Pacific Northwest National Lab, Richland, WA

Separation of non-volatile reaction products in ionic liquid catalytic processes has been a challenging problem yet is crucial for practicality, due to high viscosity, non-volatility, and easy solidification of the ionic liquid fluid. In this presentation, we'll discuss separation of 5-hydroxymethylfurfural-type (HMF) products produced from catalytic conversion of sugars in an ionic liquid fluid. HMF is regarded of large potential as a polymer monomer to replace the petroleum oil-derived polyester. The ionic liquid catalytic process provides fairly highly activity and HMF yield (>90%) compared to other solvent-based or solid catalyst-based processes. Extraction of HMF by use of an organic solvent is associated with low solubility and selectivity, which would result in complex process flow diagrams and high cost. Several solid adsorbents are identified in this work to provide high adsorption capacity (>10 wt. %) for selective adsorption of HMF from an ionic liquid/water reaction mixture. The saturated adsorbent can be readily regenerated by water and solvent washing, and the adsorbent shows stable performances over repeated adsorption/regeneration cycles. Production of 97% pure HMF is shown with a zeolite adsorbent-packed adsorption column. The adsorption can be an effective separation process technology for ionic liquid catalytic processes, where separation by changing temperatures (distillation or crystallization) is not feasible.

See more of this Session: Novel Catalytic and Separation Process Based On Ionic Liquids
See more of this Group/Topical: Topical G: Innovations of Green Process Engineering for Sustainable Energy and Environment

(295g) Membrane Extraction of Reaction Product From Ionic Liquid Reaction Process

Jian Zhang, Pacific Northwest National Lab, Richalnd, WA and Wei Liu, Energy and Environmental Directorate, Pacific Northwest National Lab, Richland, WA

In this paper, a membrane reactor is studied for continuous extraction of 5-hydroxymethylfurfural (HMF) from catalytic conversion of biomass using ionic liquid. The membrane is made of the silicalite zeolite material supported a thin, porous metal sheet, which provides chemical and thermal stability to conduct extraction under reaction conditions. The silicalite zeolite provides selective adsorption of HMF from an ionic liquid reaction mixture so that HMF is transported across the membrane into a volatile organic solvent. The HMF is recovered by evaporation of the solvent from the extract.

See more of this Session: Novel Catalytic and Separation Process Based On Ionic Liquids
See more of this Group/Topical: Topical G: Innovations of Green Process Engineering for Sustainable Energy and Environment

(335a) Membranes Technologies for Chemical Processes to Produce Clean Water and Energy, Biofuel and Pharmaceutics

T.S. Chung, Chemical and Biomolecular Engineering, National University of Singapore, Singapore, Singapore

Clean energy and water shortage, affordable healthcare and global warming are four major concerns globally resulting from resource depletion, record high oil prices, clean water shortages, high costs of pharmaceuticals, and changing climate conditions. Among many potential solutions, advance in membrane technology is one of the most direct, effective and feasible approaches to solve these sophisticated issues for chemical industries. Membrane technology is a fully integrated science and engineering which consists of materials science and engineering, chemistry and chemical engineering, separation and purification phenomena, statistical mechanics-based molecular simulation, process and product design. We will cover several areas such as:

1. The basic principle of membrane formation 2. Membranes in chemical processes for energy development (CH4, H2 and biofuel) and CO2 capture 3. Membranes in chemical processes for emerging desalination technologies: forward osmosis and membrane distillation 4. Membranes for pharmaceutical separation and life science

Email: chencts@nus.edu.sg Websites: http://www.chee.nus.edu.sg/people/faculty_chungneal.html http://www.chbe.nus.edu.sg/membrane

See more of this Session: High-Throughput and Compact Chemical Processing Technologies
See more of this Group/Topical: Topical G: Innovations of Green Process Engineering for Sustainable Energy and Environment

(335b) Microreactor for Industrial Applcation: Technical and Economical Benefits

Sergio Pissavini, Corning SAS, Avon, France

A glass based high throughput microreactor technology is presented illustrating the basic design window, technical data and configuration. An overview of some class of reactions performance is provided as compared to batch process. For three typical reaction a comparison between typical plant design batch versus micro-reactor is presented with an economical and footprint analysis comparison. The econmical and technical advantage are then projected versus their "green" impact.

See more of this Session: High-Throughput and Compact Chemical Processing Technologies
See more of this Group/Topical: Topical G: Innovations of Green Process Engineering for Sustainable Energy and Environment

DESIGN ENHANCEMENTS OF AN HIGH TEMPERATURE, CERAMIC μ-CHANNEL HEAT EXCHANGER

M. A. Wilson, J. N. Cutts and J. R. Fellows

Ceramatec, Inc.

Salt Lake City, UT

ABSTRACT

The recent developments in the energy and process industries have lead to more efficient, higher-temperature processes and compact equipment. By using small features such as micro-channels, pin fins or plate fins the length scale for which transport occurs is minimized and thus increases the heat transfer. By integrating high-thermal conductivity ceramic foams into these compact heat exchanger designs, even greater benefits of reduced hydraulic diameters and increased surface areas can be realized in environmentally durable, structurally strong materials. In this paper we will discuss several design configurations coupled with the inherent properties of preferred ceramic materials to assess the effects of foam pore size and overall porosity. These foams have been laminated into micro-channel assemblies and co-fired into monolithic devices for operational testing. The interfacial thermal-contact resistances between the porous foams and the dense films were eliminated by a co-firing process. Within these studies hydrodynamic and heat transfer models were used to predict the performance and viability of next generation high temperature heat exchangers.

INTRODUCTION

Open-cell macro-porous silicon carbide (SiC) foams, which provide high surface area to mass ratio, are well suited for heat exchange (HX) designs requiring good corrosion resistance and strengths, at elevated temperatures[1], while also providing excellent thermal conductivity. Strength and permeability, however, are among several considerations that are required in the proper selection of SiC foams used in any particular HX design. The characterization of strength or permeability is often a function of the morphology of a given foam structure, with differing porosities, strut configurations and associated pore per inch (ppi) values. To make proper choices in the selection of a foam used within an HX device, it is important to understand the correlation between the structural morphology and the attendant effects on strength and flow characteristics. Several SiC foams, processed using a replica technique, and designated initially by ppi values, have been processed and prepared to study how the morphology of the sintered foam impacts the strength of the foam body under compressive loads, as well as how the flow rates vary as a function of pore distribution and porosity values.

Figure 1 illustrates how these porous structures can enhance the heat transfer performance in a gas to liquid heat exchanger. For example, fluid channels are small for the relative high-density, high-heat transfer characteristics and low pumping power required for the fluid. The foam side (or gas side) of the heat exchanger provides high surface area and a small length-scale to enhance the heat transfer. Heat is rejected to the micro-channel walls where it is conducted through the porous matrix into the gas-side foam matrix. The cross-section of the gas-side channels is large due to its low-density, low-heat transfer characteristics and low pressure drop constraints. The gas-side integrates "extended fins" throughout the channel increasing the surface area from which heat is convected and reduces the length-scale between "cool gas" and "hot walls". This foam increases this heat transfer area from the traditional x-y planar surface to the surfaces of fibrous struts and membranes reaching into the 3rd dimension that make up the porous body.

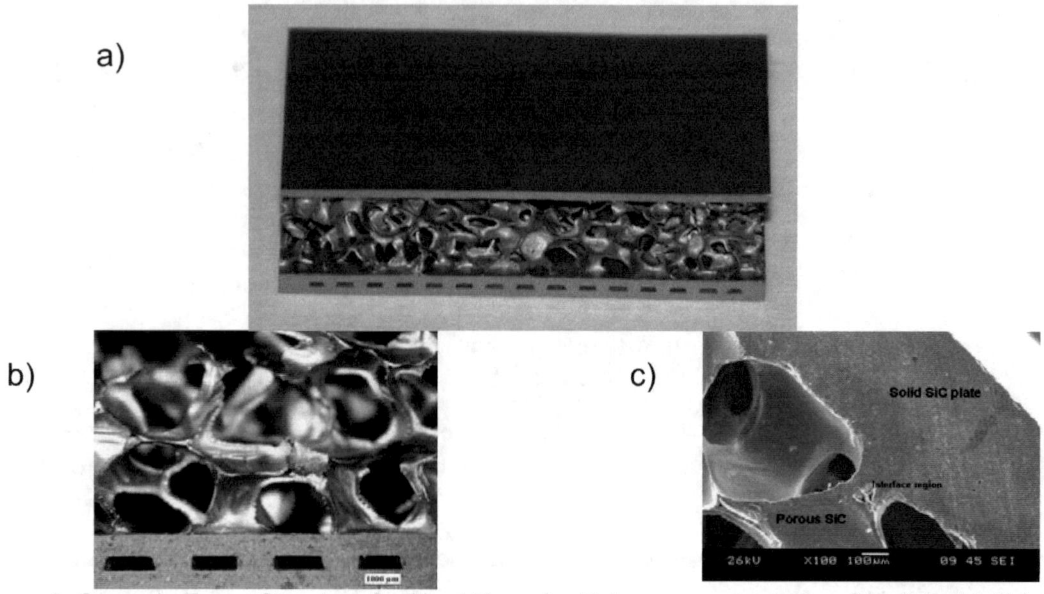

Figure 1. Ceramic Foam Structure for Heat Transfer Enhancement, a) Gas-Side/Liquid-Side Repeat Unit, b) Optical Micrograph Showing Detail of Porous Structure, c) SEM Micrograph of Porous/Dense Interface

The scope of this research was to develop porous ceramic structures and to evaluate their functionality in heat exchanger applications. In its entirety, this research will include the evaluation of the following characteristics of these foams and provide engineering correlations enabling the design and optimization of foam structures for specific applications. However, since this research is incomplete the characteristics to be discussed in this paper are indicated below:

1) Mechanical Strength – is characterized by the compressive strength of porous foam.
2) Permeability – is characterized by the pressure drop using near ambient air.
3) Net Heat Transfer Performance – Preliminary evaluations were completed wherein the heat transfer and pressure drop in foam structures were measured and are reported.

EXPERIMENTAL PROCEDURE

Various foam samples were machined into heat transfer evaluation coupons (1" x 1" x 0.25"). These foam samples were made by multiple vendors (Ceramatec, Ultramet, Duocell, Poco and Vesuvius) and represented multiple materials (silicon carbide, carbon, and aluminum).

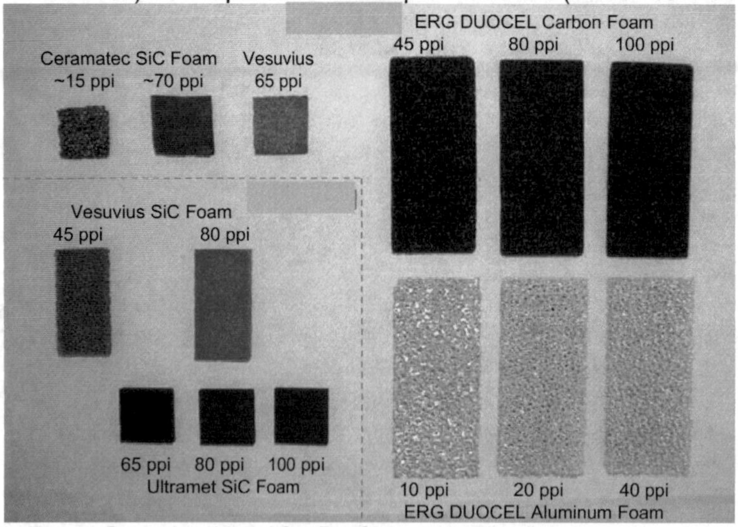

Figure 2. Foam Samples Used for Performance Testing

The heat transfer evaluation tests were configured such that the largest surfaces (1" x 1") were in thermal and mechanical contact with copper blocks to maintain isothermal surfaces, and air flowed through the porous structure in a cross-wise manner as seen in Figure 3. Using this apparatus, the pressure drop and the gas stream temperature change could be measured. Additionally, by measuring the temperature gradient in the copper block the heat flux and the temperature at the copper/foam interface could be interpolated. Given these temperatures and flow rates, an energy balance could be used to assess the net heat transfer characteristics of the porous foam structures.

Subsequently, other foam samples were fabricated to better assess the mechanistic behavior of the foam structures with respect to its mechanical strength and gas permeability. These foams were characterized in terms of their geometric and physical parameters; these included: porosity and pore size.

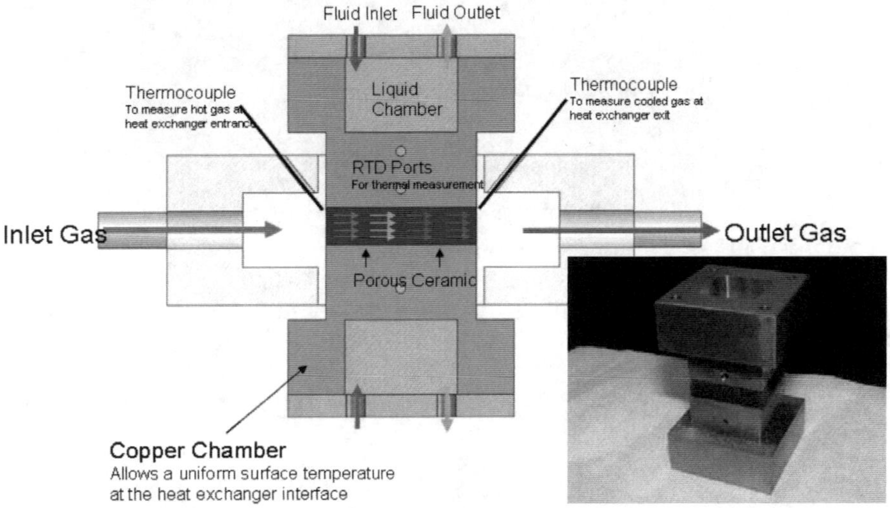

Figure 3. Heat Transfer Performance Testing of Foam Structures

Processing of ceramic foams, as described by Studart *et al.*[1] was accomplished by utilizing the replica technique, with synthetic templates. Upon dip-coating synthetic templates, samples underwent organic burnout, with subsequently sintered. Using commercial foams designated by pores per inch (ppi) designation, ten columns of 19 mm diameter by 57 mm length of sintered SiC foam were fabricated for each ppi designation (20, 30, 60, 80, and 100 ppi). Figure 4 shows optical photographs of typical sintered foam from each of the group.

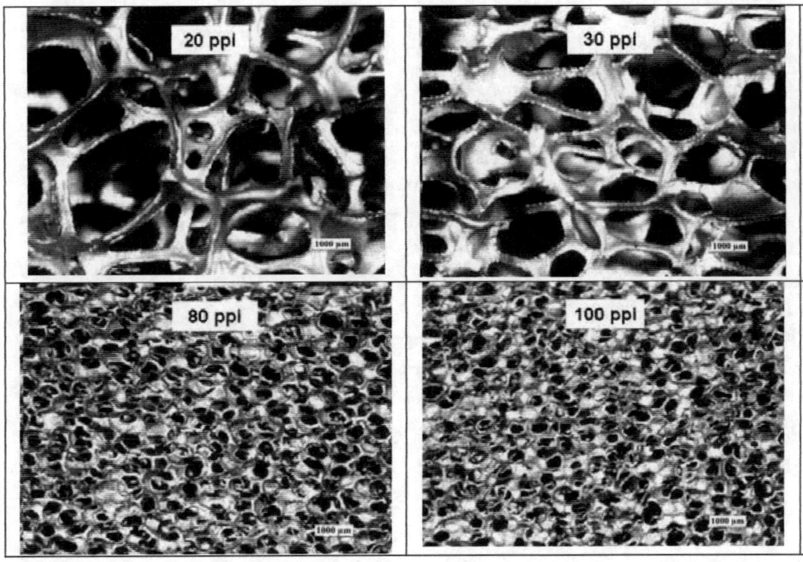

Figure 4. Optical photographs of various sintered SiC foams used in this study.

Within each group of ten, five specimens were tested in compression for strength measurements, while the other five where subjected to flow tests to determine permeability characteristics. All compression specimens were loaded in compression with a compliant foam layer was added to the top and bottom of each column to minimize uneven point loads. A typical compression test set-up is seen in Figure 5, with 60 ppi SiC foam being tested.

Figure 5. Compression test set-up.

Flow tests were conducted, with air at room temperature, by sealing individual specimens within a heat shrink wrap and flowing air at various flow rates from 5 to 20 standard liters per minute (slpm) through these columns as seen in Figure 6.

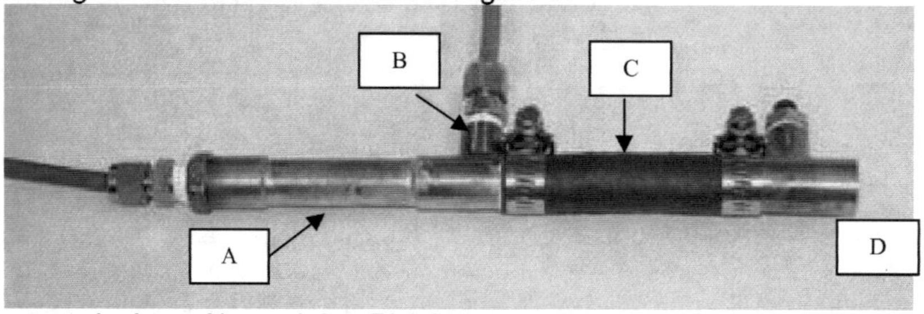

Figure 6. Flow test device. A) gas inlet, B) inlet pressure port, C) shrink wrapped SiC foam, D) gas outlet

RESULTS AND DISCUSSION

HEAT TRANSFER EVALUATIONS

These preliminary tests measured the heat transfer and pressure drop performance data of foam structures. The heat transfer performance is computed as the heat transferred (mass flow rate * Cp *ΔT of air) divided by the log mean temperature difference between the gas and constant wall interfaces. The pressure drop was measured from the inlet stream to the outlet. It was seen in the heat transfer and pressure drop data, different pore structures (manufacturer's products) even with similar PPI values have very different performance characteristics (Figure 7). It was apparent that the traditional pores-per-inch classification of these foams is insufficient for predicting their permeability. In fact, the 80 ppi foam manufactured by Vesuvius had a porosity of 58% while the Ultramet and Ceramatec 80 ppi foams had porosities between 80% and 90%. Optically, the major difference between the Ultramet and Ceramatec foams was the characteristics of the "windows" connecting adjoining cells whereby gas must pass through to permeate from upstream to downstream. Further characterization of these windows may include diameter or area and percent blockage.

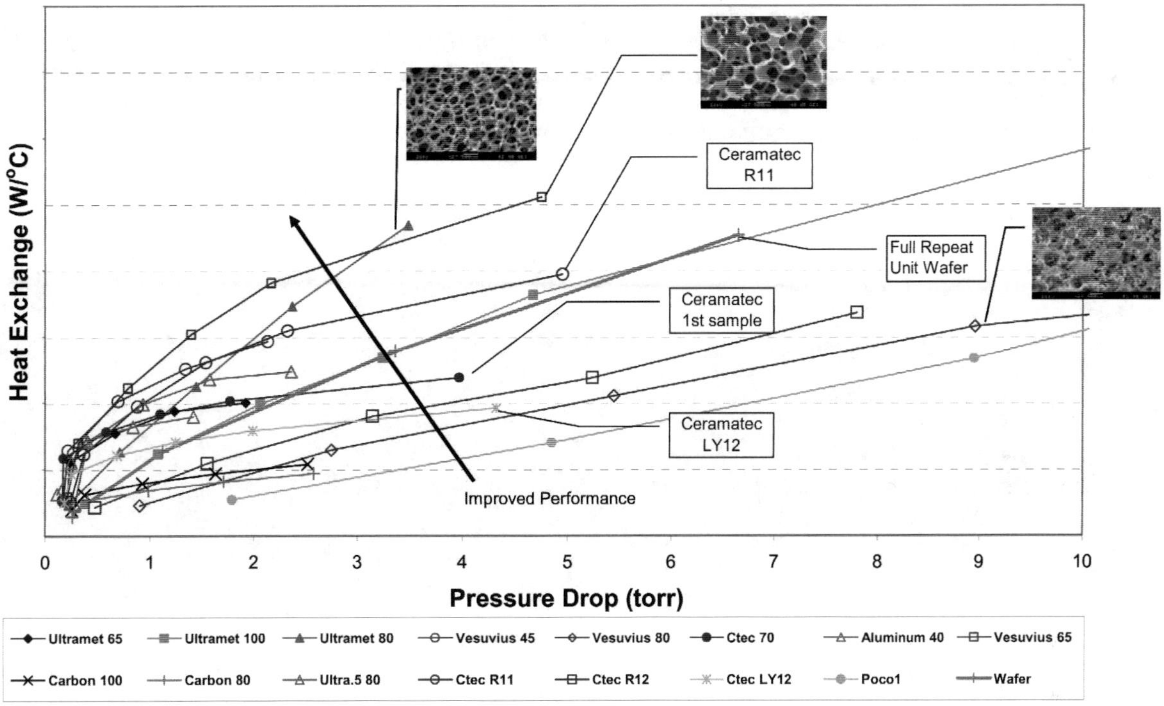

Figure 7. Relative Performance of Foam Structures

Ultimately, the objective of this research is to develop heat transfer correlations and pressure drop correlations such that foam-based heat exchangers can be optimized and compared to other compact heat exchanger designs for design selection. Kays and London[2] have compiled the performance of many compact heat exchangers for design comparisons. Appropriately, these comparisons are made using traditional non-dimensional variables (scaling parameters) such as the Reynolds number, Stanton number, Prandtl number and the friction factor. In order to define these scaling parameters, local geometric dimensions must be characterized. This lead to the second round of samples to better characterize the mechanisms influencing the performance of these foam structures as related to heat exchangers.

COMPRESSION RESULTS

The geometrical and morphological characteristics as well as the strengths for each group of sintered foam are averaged and tabulated in Table I. Upon crushing the foams under a compressive load, data was tabulated to determine the stress (σ^*) at which the cylindrical foam structure failed.

Table I. Morphological Parameters Characterizing Porous Foam Samples with Associated Average Strengths and Strength Constants.

Group PPI	Sintered PPI	Porosity (%)	Stress σ^* (MPa)
20	19.3	93.5	0.85
30	28.0	93.0	0.91
60	66.1	77.1	1.58
80	83.9	89.6	1.34
100	95.6	88.0	1.17

Data for all specimens tested in compression was tabulated, with the relationships of strength (σ*) to porosity and ppi shown in Figure 8. Here it can be seen that strength would not correlate well with pore spacing. However when comparing strength as a function of porosity, we find, that strength increases nonlinearly as porosity decreases, as would be expected[3]. Hence, porosity is a more predictive parameter in determining strength of the foam than the pore spacing.

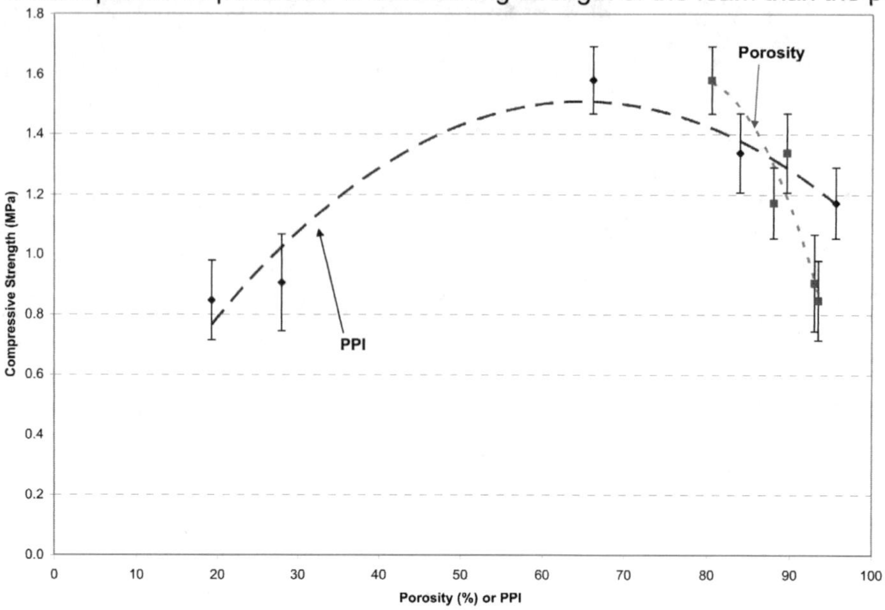

Figure 8. Strength as a function of porosity (%) and PPI.

PERMEABILITY RESULTS

Figure 9 shows the general trend of permeability for all samples tested, with pressure drop (normalized) increasing as porosity decreases. As also indicated in Figure 9, the pressure drop seems to align with inverse porosity and foam ppi designation.

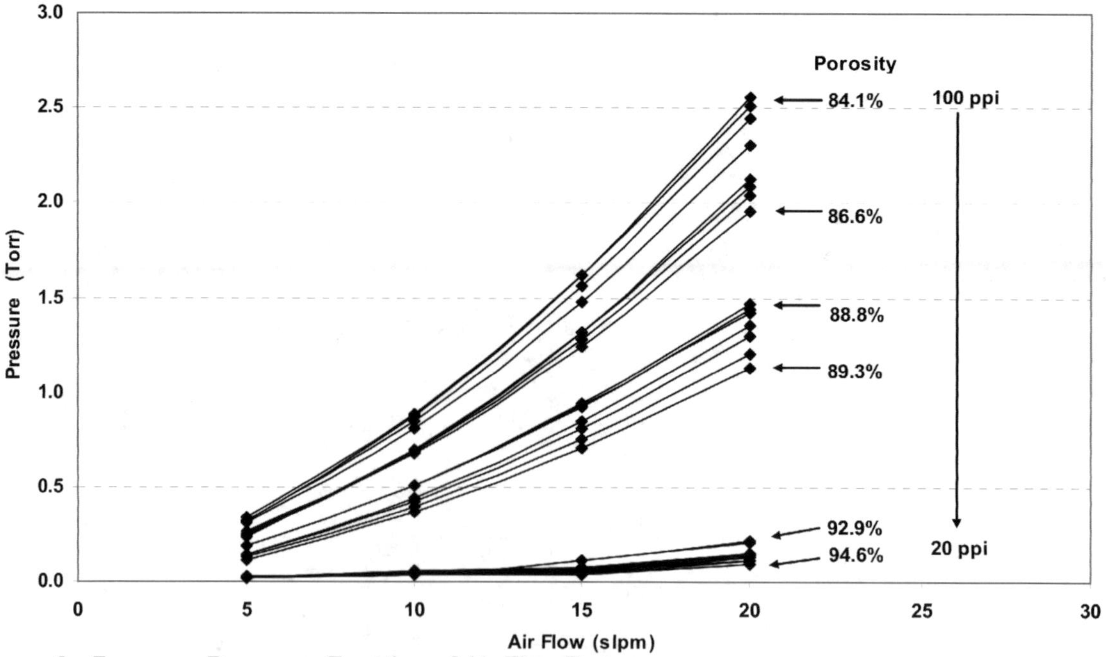

Figure 9. Pressure Drop as a Function of Air Flow Rate.

The change in pressure, or pressure drop ΔP, is a function of several variables given by[4]

$$\Delta P = f\left(\frac{l}{D}\right)\rho\frac{V^2}{2} \qquad (1)$$

where f is the friction factor, l is the tube length, D is the tube diameter, ρ is density of the gas that flows through the tube, and V is the free stream velocity of the gas.

As discussed previously, the window area between cells may limit and determine the flow characteristics of the foam. It is well know that for laminar internal flows the friction factor (f) is inversely proportional to the Reynolds number (Re)[7]. For example in circular, unblocked pipes the relationship becomes f=64/Re; for high aspect ratio rectangular, unblocked ducts this is f=96/Re[5]. By assembling and reducing this pressure drop data for flow through porous foams this relationship is similar, yielding a proportionality constant of about 156 as seen in Figure 10, based on the dimensions of the cell to cell windows. This last correlation shows that to best tailor foam to meet specified pressure drop performance, the window characteristic of diameter is of primary importance. It should be noted that the dimension of these windows will vary with pore spacing and porosity. In other words for a preferred window area, there is range of porosities that can be attained for a given pore spacing of the synthetic template, and vice versa. Therefore, once the target flow and pressure drop have been determined, the necessary window diameter for the foam can be engineered and constructed using a specified range of pore openings per unit length and/or porosity characteristics.

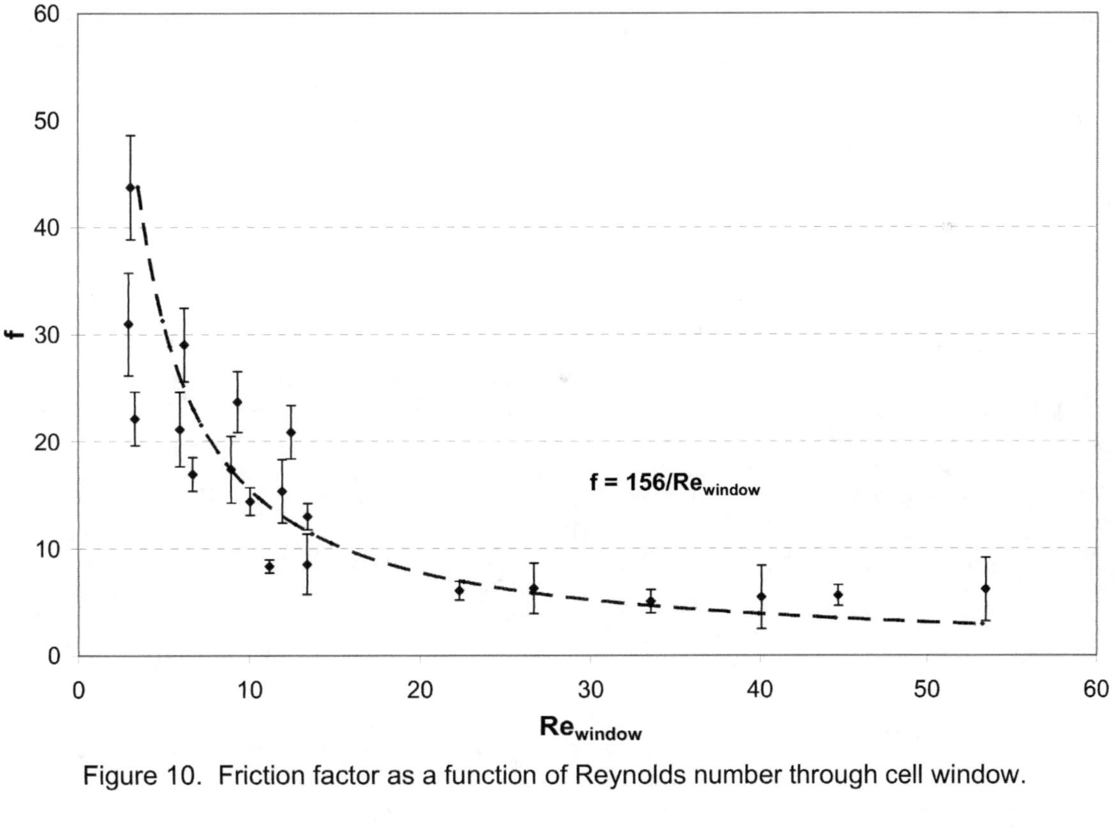

Figure 10. Friction factor as a function of Reynolds number through cell window.

SUMMARY AND CONCLUSIONS

SiC foam is an ideal candidate material for HX systems due to its thermal properties and strength. This work shows how to better quantify SiC foams produced using the replica method and synthetic templates.

The strength of the SiC foams tested are strongly dependent on porosity rather than pore spacing. As porosity decreases there is an increase in strength under compressive loading, with a maximum load of 1.6 MPa, correlating to a porosity of 77%.

Further, flow tests of these foams have shown that pressure drop and friction factor are functions of both porosity and ppi. These data and correlations have shown that the relationship between the friction factor and the Reynolds number of the window between cells are of primary importance in determining flow characteristics and subsequent tailoring of foams.

By examining several foams structures and materials it was found that improved heat transfer performance can be attained by modifying the morphology of the porous structure. One metric in characterizing foams is the pore size (typically PPI – pores per inch) another is the porosity (% theoretical density). Typically, moderately high porosity foams perform best. Foams that have porosities in the 90^{+}% range have reduced surface areas and ligament cross-sectional areas, thus possibly degrading their capability to transfer heat from a thermal substrate to a flowing stream. Foams that have moderate porosities (less than about 60%) have excessive flow blockage and hence minimize the flow rates of the gas stream to be convectively heated or cooled. The optimum appears to be within the 70% to 90% porosities range. Further optimization will largely be determined by the micro-structure features of the foams (strut diameter, window blockage etc).

ACKNOWLEDGMENTS

This work was initiated under a Department of the Air Force SBIR contract FA8650-08-2821. Subsequent follow-on work was done under the internal sponsorship of Ceramatec, Inc.

REFERENCES

---

[1] A. R. Studart, U. T. Gonzenbach, E. Tervoort, and L. J. Gauckler, "Processing Routes to Macroporous Ceramics: A Review," *J. Am. Ceram. Soc*, 89[6] 1771-89 (2006).
[2] Kays, W. M. and London, A. L.; "Compact Heat Exchangers", Reprint Edition 1998, Krieger Publishing Company, Malabar, FL, 32950.
[3] R. W. Rice, "Porosity of Ceramics" (1998), CRC Press.
[4] Alan Mironer; "Engineering Fluid Mechanics", 1979, McGraw-Hill, Inc.
[5] J. P. Holman, "Heat Transfer", 5th Ed., (1981), McGraw Hill Book Co.

TOWARD LOW COST FABRICATION OF MICROCHANNEL PROCESS TECHNOLOGIES – COST MODELING FOR MANUFACTURING DEVELOPMENT

Steven D. Leith and Dale A. King, Pacific Northwest National Laboratory, Richland, WA
Brian K. Paul, Oregon State University, Corvallis, OR

Introduction

Chemical and energy conversion systems based on microchannel process technology (MPT) demonstrate high performance in applications in which rates are controlled by diffusive heat and mass transfer flux. MPT-based heat exchangers, absorbers/desorbers and chemical reactors all benefit from process intensification and have been used in a variety of mobile energy conversion systems including fuel reformers/converters, heat pumps and waste heat scavenging technologies. The service environments typical of MPTs often require the devices to be fabricated from metals such as aluminum, titanium, stainless steel or high temperature super alloys. Flow channels and associated critical dimensions in these devices can be as small as 50 um, but generally range from 100 to 1000 um in width and height with characteristic flow channel lengths normally in the mm to cm range. High surface area architectures (*e.g.* wicks or textured surfaces) are often included in the flow channels as well for enhanced mass transfer and/or catalytic functionality.

Heat activated cooling systems that utilize MPT devices are being developed at the Microproducts Breakthrough Institute (MBI) for applications in which reduced system size and weight are of high value and/or electricity generation costs are high (*e.g.* mobile space cooling units in forward deployed military field operations). A typical system is illustrated in Figure 1 and is based on an aqueous ammonia absorption/desorption cycle driven by an external heat source. In this particular system design, the desorber, absorber and heat exchanger represented schematically in (A) are MPT devices and are integrated with off-the-shelf components in the heat pump system (B). The absorber device (C) is the largest MPT component in the system and is used to regenerate concentrated ammonia solution by contacting ammonia vapor with a weak aqueous ammonia solution in a high surface area mixing architecture.

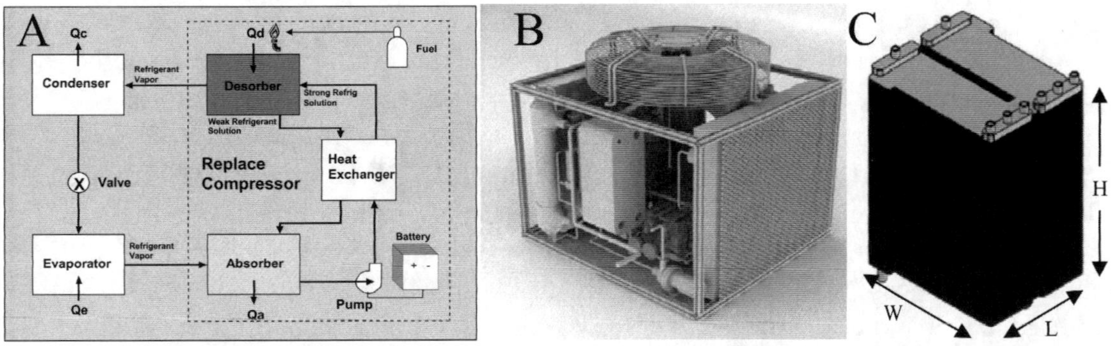

Figure 1. Ammonia absorption cycle heat pump system (A and B) utilizing MPT devices. The 5 kW system requires two absorbers (C) each with external dimensions of approximately 15 cm (W) x 15 cm (L) x 25 cm (H).

MPT enabled heat pumps demonstrate high COP for a given system size, are quiet in operation and can be driven by renewable energy, combustion of natural gas or, more notably, by waste heat from an internal combustion engine or fuel cell. This characteristic energy use flexibility, unit mobility and relatively small system size make MPT heat activated cooling systems performance- and cost-competitive with existing technologies in a number of military and commercial markets. To move from technology demonstration to market entry and commercialization, however, requires an better understanding of the manufacturing processes and associated cost structures necessary to fabricate components in production volumes. Recent work at the MBI has focused on developing volume-sensitive cost estimation models for predicting the manufacturing costs of MPT devices fabricated using a range of processing technologies. The process-based cost models are used to develop an understanding of the economic trade-offs between candidate processes and are utilized in a design for manufacturing approach to MPT device development. In this paper we present results and analysis of the cost modeling effort to date and apply the methodology in a case study of the heat pump absorber device designed, built and tested by Pacific Northwest National Laboratory.

Process-Based Cost Modeling

Process-based cost models have been used to estimate manufacturing costs for a range of materials and products including diamond films, carbon nanotubes, steel powder and injection molded plastics (1-4). Using this approach, the total manufacturing cost is broken down into a number of cost elements associated with each step in the production process. Each cost element is estimated separately and then all are summed to establish an overall manufacturing cost. Typical cost elements include capital equipment, labor, direct materials, indirect materials, energy/utilities, facilities and maintenance. Variation in each cost element is estimated through creation of functional relationships that quantify the impact of manufacturing variables (*e.g.* production volume, equipment through-put, labor and loading rates, electricity cost, etc.) on each cost element. The output from the process-based model is an estimate of cost of goods sold (COGS); typical overhead cost contributors such as sales and marketing, R&D, administration, management, profit and taxes are not considered. Application of the process-based cost modeling methodology to manufacturing of a device such as the absorber shown in Figure 1 can be used to illustrate the relative contribution of design, materials and fabrication process choice on final device cost. Identification of these cost drivers is a critical enabler in effectively prioritizing the fabrication processes that give the highest probability for commercial success. Cost model sensitivity analyses can be used to determine areas of technical and/or financial risk and are a useful tool in developing more sophisticated economic assessments; for example supporting a "make vs. buy" decision in manufacturing production. The process-based cost model is used here to identify primary cost drivers in the manufacture of the ammonia absorber device and to illustrate the interaction of design and COGS across a range of production volumes.

Case Study – Manufacturing Cost of MPT Absorber Device

The first step in the modeling methodology is to define the fabrication process and develop the functions and equations that relate manufacturing parameters to the individual cost elements. A typical fabrication process flow sheet for manufacturing of MPT devices is illustrated in Figure 2. The architecture of the absorber device is based on a stacked-shim fabrication process in which shims are first patterned (with flow channels and/or fluidic vias) and then bonded in a monolithic stack. Two specially machined end-plates are joined to either end of the device to allow addition of interconnects for fluidic communication to and from the device. The service environment of the MPT device has a great influence on the choice of shim and end-plate materials and, in turn, the material choice has a significant impact on available fabrication processes and processing conditions. In the present study, the service environment of the absorber requires mechanical strength at elevated temperature in corrosive fluids. To meet the material performance requirements of the prototype system, austenitic 316 stainless steel is used for both shim and end-plate architectures.

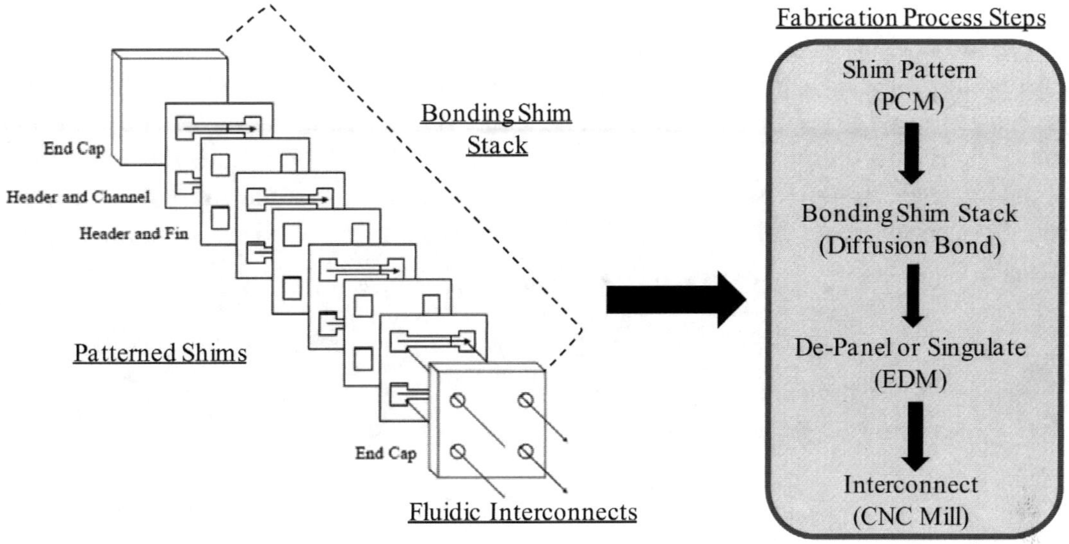

Figure 2. Typical construction scheme and fabrication process steps for stainless steel MPT device.

A number of fabrication processes have been used at MBI to build demonstration devices. The most-widely used process steps, however, appear in Figure 2 and are especially well-suited for fabricating metallic-based MPT devices at low (prototype) volumes. In the present study, this "default" manufacturing process flow is considered in which photochemical machining (PCM) is used to pattern shims, diffusion bonding for joining of the shim stack, electro-discharge machining (EDM) for final device perimeter shaping and CNC milling to define fluidic interconnects.

At demonstration and prototyping fabrication volumes, the cost associated with shim patterning and shim stack bonding has been a significant fraction of the total fab cost for MPT devices made at MBI. Typically, however, these fabrication processes have been performed by third party contract manufacturers and the cost of device fab is very likely not representative of that which could be achieved using a dedicated production line as is assumed with the cost model. Using the default process flow illustrated in Figure 2, the process-based cost model is used to predict manufacturing costs of the absorber over a range of production volumes representative of the anticipated market. Figure 3 illustrates the cost breakdown for manufacturing between 100 and 10,000 absorber units/year. Consistent with past prototyping efforts, at low volumes (100/yr) the majority (80%) of the manufacturing cost is contributed by the PCM shim patterning and diffusion bonding processes. At these volumes, the direct material cost of the stainless steel shims and end plates accounts for nearly all the remaining cost (15% of total) while the EDM and CNC milling processes contribute very little. As manufacturing volumes increase, however, substantial reductions in device cost are realized. In fact, at fairly modest production volumes of 1,000 to 10,000 absorber units annually, the unit COGS is estimated to be between \$3,000 and \$4,000. Consistent across all absorber production volumes is the cost contribution of the PCM process. Model simulations predict that PCM costs actually increase as a fraction of the total with increased volume production even while decreasing on a unit COGS basis.

The cost model is used further to explore the cost drivers inherent in the PCM process and to evaluate the financial viability of this option as a candidate fabrication option in an MPT production environment. Figure 4 illustrates the area-based production cost of using PCM to pattern stainless steel shims using an industry standard etching line designed for a maximum capacity of 40,000 m^2/yr. When the direct material cost (*i.e.* 316 SS shim stock) is removed from the analysis, the cost model predicts a decrease in manufacturing cost from approximately \$275/m^2 at low production volumes to a minimum of about \$50/m^2 at production volumes which approach maximum design capacity. For the absorber architecture, the \$50/m^2 COGS estimate is equivalent to approximately \$1.25 per shim. The rate of cost reduction with volume predicted for the PCM process is characteristic of a manufacturing process line designed for a specific production volume. At low production, the line is under-utilized and fixed costs associated with capital and labor are allocated to a small number of parts, driving the unit cost up. As production volumes increases, both capital and labor utilization improve and unit COGS decrease.

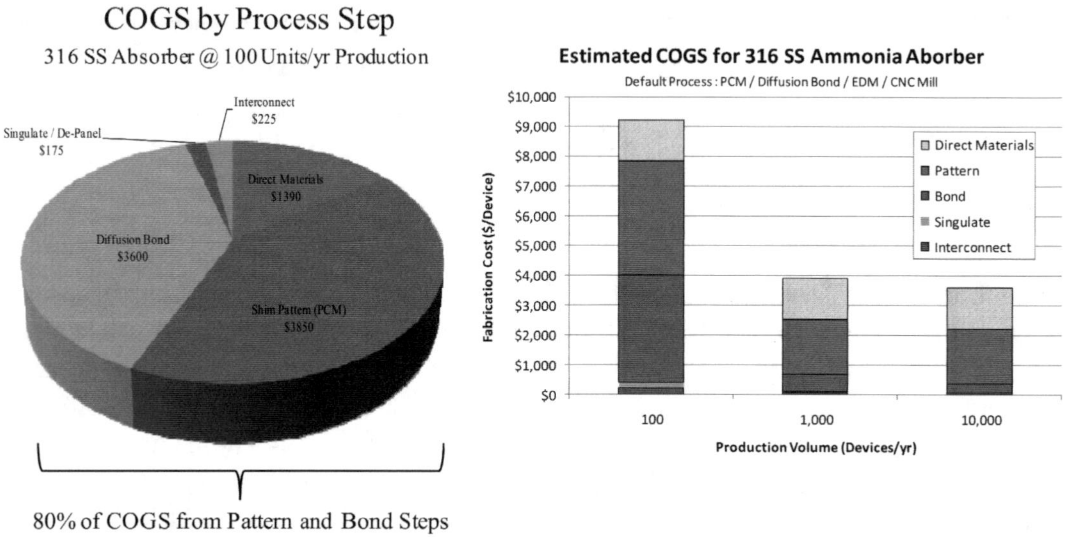

Figure 3. Manufacturing cost breakdown for fabrication of absorber device at 100, 1000 and 10,000 units/yr production.

Similar cost estimation results and breakdowns to those illustrated in Figure 4 can be obtained by applying the cost model methodology to other candidate processes of interest. For example, shim patterning can be performed using a number of other fabrication processes including electrochemical machining (ECM), stamping (piercing or coining), laser cutting and abrasive water jet cutting. Direct comparison of the cost associated with each of these processes for the patterning of specific shim architectures illustrates which candidate processes are likely to be the most economical in production. In this way, alternative shim designs can be considered that might enable a lower cost fabrication path.

Figure 5 summarizes the process costs for patterning shims with fluidic vias (A) and "blind" flow channels (B) representative of architectures required for the absorber (*i.e.* 500 um thick, 316 SS shims). Patterning of fluidic vias can be and has been performed using each of the patterning processes listed above. One design attribute that can have a significant impact on pattern cost is the density of vias on a given shim. For serial patterning processes (*i.e.* those that raster a point energy source across the shim surface) such as laser cutting and abrasive water jet, the via perimeter per unit area directly influences process through-put and, ultimately, cost. Alternatively, for parallel patterning processes (*i.e.* those which pattern the entire area at once) such as PCM and ECM, the intrinsic processing rate and final cost is independent of via pattern density. The piercing process, however, has characteristics similar to both parallel and serial processing and exhibits a modest sensitivity to via perimeter area. For consistency in comparison, the COGS estimates summarized in Figure 5 (A) are based on a shim architecture with a via perimeter of 180 m/m^2.

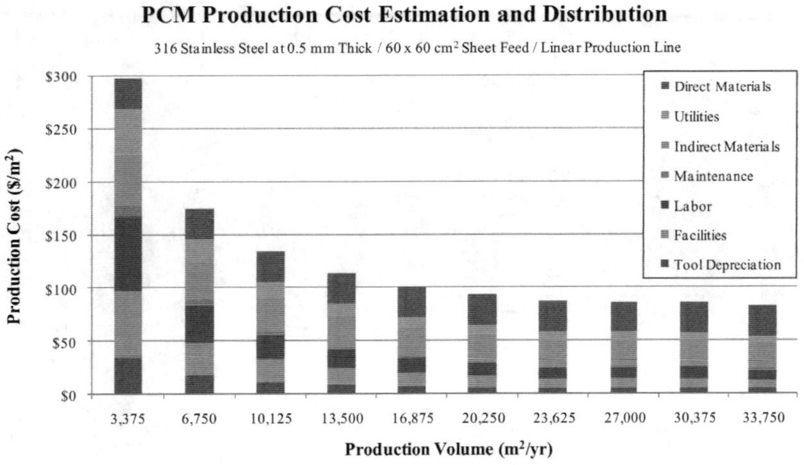

Figure 4. Estimated manufacturing cost for patterning shims using PCM

The summary of estimated costs for patterning fluidic vias in stainless steel sheet stock using the different candidate processes illustrates the volume sensitivity and relative costs associated with each option. At low production volumes, manufacturing costs increase for all process, but most sharply for PCM processing. At volumes of 24,000 m^2/yr and greater, however, the estimated COGS for all options except abrasive water jet fall between \$50 and \$75/m^2 (approximately \$1.25 and \$1.75 per shim for the absorber). This cost differential is modest, but can be significant for devices that require a high number of shims. The final choice of patterning process, therefore, will likely require consideration of process capability (e.g. minimum feature size, edge burr, sidewall profile, etc.) vs. pattern dimension tolerances.

Figure 5 (B) summarizes the cost of patterning flow channels, or blind features, in an MPT shim using the different candidate processes. Note that only three of the five patterning processes are included; the capability of both laser and abrasive water jet patterning of flow channels is questionable and not yet production ready. Coining of flow channels is also a challenging process and puts specific limitations on channel design. This approach, however, has been demonstrated for production of fuel cell components and given the similarity to coining MPT architectures is included in the analysis. Similar cost reduction with volume as with via patterning is predicted by the model; in fact the cost estimations for PCM and ECM are essentially identical for patterning both vias and flow channels. Both of these processes are capable of creating vias and flow channels in a single production run, a distinct advantage over alternative processes and a significant reason why PCM has been and looks to continue as a primary option for shim patterning in MPT manufacturing.

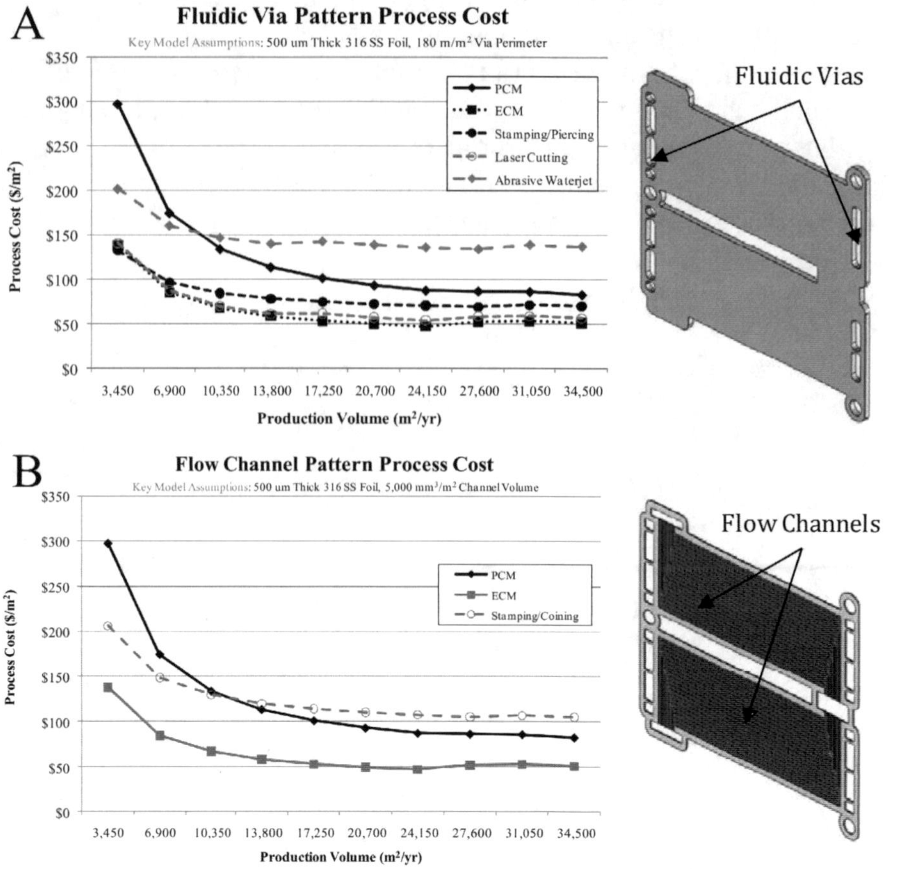

Figure 5. Cost comparisons for patterning shims using different candidate processes.

As illustrated in Figure 3, the bonding of shim stacks also contributes significantly to the overall manufacturing cost of MPT devices. In Figure 6, the cost of two candidate bonding processes are compared; diffusion bonding and laser welding. As with the serial patterning processes discussed above, laser welding cost is dependent on the total weld joint length in the device. As shown in Figure 6 (A), welding lengths between 6,000 and 120,000 cm per device are considered. Alternatively, diffusion bonding is not dependent on weld joint length, but rather on the number of devices bonded in a single process cycle. Two device through-puts are included in the cost estimation results for diffusion bonding summarized in Figure 6. The solid line simulates a bonding process in which four devices are bonded in each cycle while the dashed line represents cost for bonding 16 devices per cycle. Consistent with increased through-put, laser welding costs are minimized for devices which require the least number of joints while diffusion bonding costs are minimized as the number of devices per bond cycle increases. The simulation results summarized in Figure 6 (A) can be used to determine which joining process is most economical for a given device design. At low volumes, the estimated cost of laser welding is lower than that predicted for diffusion bonding, a direct result of the high capital cost associated with industrial scale diffusion bonding equipment. At higher volumes, however, the choice is not completely clear and will likely depend on process capability (*i.e.* yield) and an assessment of technical risk (*i.e.* the likelihood that bonding 16 devices simultaneously is feasible). The cost model can be used to explore the impact of yield through sensitivity analyses, but the assessment of technical risk and impact on capital investment choice is ultimately a business decision.

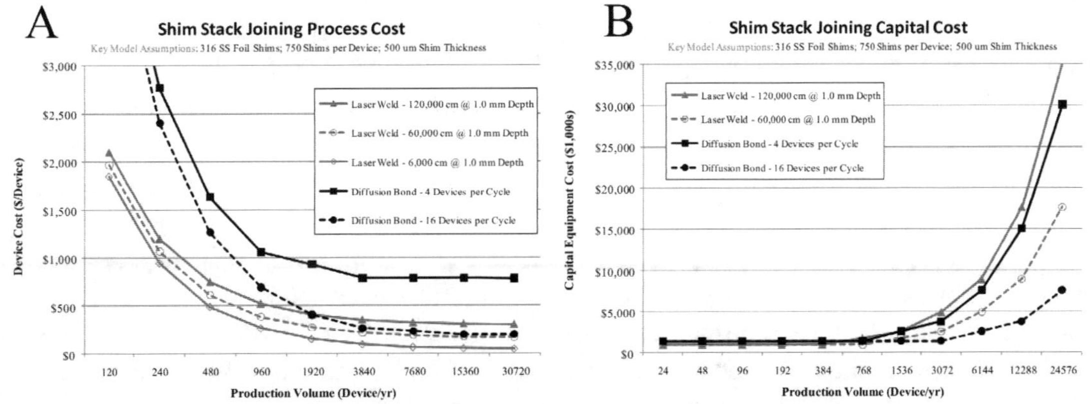

Figure 6. Shim stack joining process cost (A) and required capital equipment investment (B).

In addition to predicting manufacturing cost for the various candidate processes, the cost model simulation is also used to estimate the required capital investment to realize a given unit COGS. For example, consider the unit COGS and capital investment estimations for laser welding vs. diffusion bonding of a device that requires 60,000 cm of laser weld joining per device. From Figure 6 (A), the lowest unit COGS at high volume appears to be same for laser welding and diffusion bonding (16 devices per cycle). As discussed above, for such a situation in which two candidate processes are estimated to have identical COGS, a logical fabrication decision might take into account the capability attributes of one process vs. the other. Figure 6 (B), however, illustrates the difference in capital investment required between the two processes. From the standpoint of financial risk, one could reasonable argue that an investment in the diffusion bonding process at $4M is preferred over a $9M laser welding investment to achieve the same unit COGS. This example illustrates an important aspect in the interpretation and use of cost model results; reducing COGS is certainly necessary for competitive manufacturing, but the capital investment required to realize the unit COGS must be considered as well. Excessive investment in capital can increase the financial risk of manufacturing and ultimately stress the commercialization effort. While the process-based cost model discussed here can provide insight into the relationship between COGS and required investment, a more sophisticated economic analysis that includes market intelligence and cost targets is required to determine what investment level is appropriate or what manufacturing cost is too high.

Summary and Next Steps

1. Process based cost models have been developed at the MBI for estimating the manufacturing cost of MPT devices using a variety of candidate fabrication processes.

2. These cost model tools are being developed and used by the MBI to estimate production costs of MPT devices to assist technology commercialization partners evaluate MPT options for processing solutions in heat transfer and reaction engineering applications.

3. Costs for absorber and absorber-like architectures at relatively modest production volumes of 10,000 per year are in the \$3000 to \$4000/unit range. This is significantly lower than contract manufacturing cost and is likely an acceptable COGS target in one or more market segments. The cost model tools will be used to facilitate design for manufacture (DFM) activities aimed at further reducing costs for similar architectures while maintaining required device functionality. Near term efforts are focused on developing market intelligence in-house and in cooperation with commercialization partners to better match the MBI MPT value proposition with market needs and cost targets.

4. Capital investment for production of absorber and absorber-like devices at relatively modest production volumes of 100 to 10,000 annual units are on the order of \$M to \$10M dollars. The cost of a dedicated PCM line at these volumes is on the order of \$1M and can be used to produce shims at a COGS as low as \$1.50 each. In addition, PCM is capable of both via and flow channel patterning, suggesting that PCM will continue to be a process of interest in MPT manufacture for the near future.

Next Steps

1. Multi-variant sensitivity analyses need to be performed to better identify and quantify technical and financial risk to commercialization. Sensitivity analyses are underway at MBI to explore impacts of design, process and materials on manufacturing cost and capital investment.

2. Model extensions to include consideration of overhead costs and total cost of ownership parameters (*e.g.* device lifetime, disposal cost, operation cost, etc.) are being developed which will further strengthen the cost understanding in more market-specific applications.

References

1. Busch, J. (1994), "Cost Modeling as a Technical Management Tool", *Research Technology Management*, Nov/Dec, pp. 50-57.
2. Isaacs, J.A., Tanwani, A., Healy, M.L., Dahlben, L.J. (2010), "Economic Assessment of Single-Walled Carbon Nanotube Processes", *Journal of Nanoparticle Research*, **12**, pp. 551-562.
3. Kirchain, R. (2001), "Cost Modeling of Materials and Manufacturing Processes", in *Encyclopedia of Material Science and Engineering*, pp. 1718-1727.
4. Chiango, D.A., Isaacs, J.A., Miller, S.A. (2000), "Production of Steel Powder by Rotating Electrode Processes: Economic Analysis", *International Journal of Powder Metallurgy*, **36**, No. 4, pp. 49-56.

(335e) A Compact Integrated Warm Syngas Overall Cleanup System

Liyu Li[1], David L. King[2], Baowei Chen[3] and Chris Howard[1], (1)Hydrocarbon Processing Group, Pacific Northwest National Laboratory, Richland, WA, (2)Pacific Northwest National Laboratory, Richland, WA, (3)Biological Sciences, Pacific Northwest National Laboratory, Richland, WA

Gasification of coal or biomass to syngas followed by catalytic synthesis of ammonia, hydrocarbons or oxygenates provides a feasible strategy to meet the increasing demand for fertilizers and transportation fuels. This syngas has many impurities, including As, Be, Cd, Cl, Cr, Hg, K, N, Na, P, Pb, S, Sb, Se, V, Zn. Most of them need to be removed down to part per billion levels (ppb) due to their strong interactions with syngas reforming and synthesis catalysts. Although technical approaches exist for removal of these species, they are rather costly, employing solvents at ambient or lower temperature and backup sacrificial sorbents. The energy efficiency of the current process would be improved if all the contaminants can be removed at temperatures higher than the syngas reforming and chemical synthesis temperatures (greater than 200oC).

In this presentation, we will describe our recent progress in developing a compact, integrated system for overall impurity cleanup from warm syngases generated from water-quenched coal or biomass gasifiers. This compact system consists of four major sub-systems: 1) a particulate filter to remove solid impurities, 2) a high capacity sacrificial absorption bed to remove HCl, 3) two sequential absorbent beds to remove sulfur from more than 1000 ppm levels down to less than 60 ppb levels, and 4) a high capacity metal-based sacrificial absorbent to remove As, P, Se, and trace level S and HCl. Each sub-system was individually developed, and a compact integrated system was built up. This integrated system was demonstrated to be able to remove all the catalysts poisons in a simulated warm syngas stream and thus effectively prevented the downstream syngas reforming catalysts from being deactivated. This compact system can also be used in stationary hydrocarbon fuel processors for clean feed gas generation for fuel cells applications.

See more of this Session: High-Throughput and Compact Chemical Processing Technologies
See more of this Group/Topical: Topical G: Innovations of Green Process Engineering for Sustainable Energy and Environment

Pyrolysis and co-pyrolysis of biomass and coal in a free-fall reactor

Ligang Wei[1,2], Shaoping Xu[1*], Li Zhang[1]

(1. State Key Laboratory of Fine Chemicals, Department of Chemical Engineering, Dalian University of Technology, Dalian 116012, China; 2. School of Chemical Engineering and Material, Dalian Polytechnic University, Dalian 116034, China)

Pyrolysis and co-pyrolysis of solid fuels, like biomass and coal, is a promising process to produce liquid fuels and chemicals. The free-fall reactor has the advantages of a higher heating rate and a short but narrowly distributed residence time (in the order of seconds) of both solids and gas. It is a good approach to investigate the mechanisms of fast pyrolysis of solid fuel in a free fall reactor. In this work, pyrolysis and co-pyrolysis of biomass and coal were performed in a free fall reactor under atmospheric pressure. Effects of type and composition of feedstock, reactor temperature, steam, and catalyst on the characteristics of the process were investigated at 500~800°C. The biomass feedstocks are agriculture residues legume straw (LS) and tobacco stalk, forestry residues pine sawdust (SD) and fruit shells apricot stone, which have different chemical compositions. Two types of coal, low rank Dayan brown coal (DY) and high rank Tiefa bituminous coal (TF) were selected as coal samples.

The experimental results show that a higher temperature favors producing the hydrogen-rich gas. Biomass composed of higher holocellulose component produces a relatively higher hydrogen-rich gas yield than that composed of higher lignin. The characteristics of pyrolysis can be inferred according to the chemical compositions of biomass samples. The presence of steam increases the gas yields and reduces the tar and char yields. It is considered that an in situ steam reforming reaction of tar and water–gas shift reaction takes place even in a short gas residence time in the free-fall reactor. Dolomite reveals a comparably good performance in terms of catalytic activity of tar destruction and the consequential increase in the production of gases in thus a short vapor-catalyst contact time.

The results of co-pyrolysis of biomass and coal shows that the char yields decrease and consequently the liquid yields increase compared with the calculated values, even the compositions of the gaseous products from the blended samples are not in accordance with those of their parent fuels, It suggests that there exist synergetic effects in co-pyrolysis of biomass and coal even in a short gas residence time in the free-fall reactor. In the co-pyrolysis of LS/DY, the obvious synergy on the products yields was observed at 600°C and the biomass blending ratio of 73 wt.%. The type of feedstock influences greatly the product yields and the CO_2 reactivity of char from the co-pyrolysis. The synergy is more significant in the presence of LS with higher holocellulose and ash content than that in the presence of SD. The liquid yield from the co-pyrolysis of high rank TF is higher than that of low rank DY. The char reactivity from co-pyrolysis in the presence of LS is much higher than that in the presence of SD, while that in the presence of TF is more active than that in the presence of DY.

The liquid products from the pyrolysis and co-pyrolysis were separated by solvent extraction-column chromatography and then analyzed by GC/MS. The results show that, the yields of asphalts and long chain aliphatic hydrocarbons from co-pyrolysis decrease in comparison to the calculated values, while the yield of light molecular weight phenol derivatives increase. The contents of decalins, which cannot be found in the tar from pyrolysis of single fuel, were 43.37% in aromatic fraction from the tar in co-pyrolysis. All these results verify that the synergy exists in the co-pyrolysis of biomass and coal in a free fall reactor. It is a contribution of the reactions such as hydropyrolysis and hydrogenation in co-pyrolysis of coal and biomass under the hydrogen atmosphere supplied by biomass pyrolysis. Co-pyrolysis of biomass and coal favor producing low molecular weight compounds and improving the quality of the liquid product.

Acknowledgements. The authors sincerely acknowledge the financial support from the National Natural Science Foundation of China under Grant No.50776013, the National High-tech R&D Program of China (863 Program) under Contract No. 2008AA05Z407 and the Program for Liaoning Excellent Talents in University under Grant No. 2009R10.

*Corresponding author. Tel.: 0086-411-39893837 , Fax: 0086-411-83683467 , E-mail : huizixu@hotmail.com

(335g) Solar Thermochemical Production of Metal Nitrides and Ammonia Utilizing Transition Metal Reactants

Ronald Michalsky, Chemical engineering, Kansas State University, Manhattan, KS and Peter H. Pfromm, Department of Chemical Engineering, Kansas State University, Manhattan, KS

One of the major products of the chemical industry, ammonia is used widely as artificial fertilizer and has been proposed recently as sustainable transportation fuel. Using transition metal-based catalysts the Haber-Bosch process produces ammonia industrially from its elements. Globally, the Haber-Bosch process consumes up to 5% of all natural gas produced and 2% of the total energy production, with significant fossil-based carbon dioxide emissions. The process cannot easily be scaled down or performed in developing countries due to the required temperature, pressure, natural gas, and utilities. Based on previous research employing Al- or Cr-based reactants to produce ammonia by a solar thermochemical reaction cycle at near atmospheric pressure and 1000-2000°C and without the need for natural gas as reactant, the work presented here studies process and reactant modifications that might allow the solar-driven ammonia production from water and air to proceed at below 1500°C, i.e., at temperatures which would not require sophisticated furnace design and construction materials. Utilizing a laboratory scale Fresnel lens-based tubular solar thermochemical reactor, formation of metal nitrides from various transition metal oxides is studied at atmospheric pressure. Liberation of ammonia from the nitrides due to corrosion with steam will be focused. The effect of ammonia formation favored at low temperatures and metal nitride corrosion favored at high temperatures is addressed. Deviation from thermodynamic calculations for bulk materials is discussed.

See more of this Session: High-Throughput and Compact Chemical Processing Technologies
See more of this Group/Topical: Topical G: Innovations of Green Process Engineering for Sustainable Energy and Environment

(380a) Engineered Metabolism for the Production of Fuels and Chemicals From Glycerol and Fatty Acids: The Role of Synthetic and Systems Biology

Ramon Gonzalez, Departments of Chemical & Biomolecular Engineering and Bioengineering, Rice University, Houston, TX

Although sugars are the primary feedstock for the biological production of renewable fuels and chemicals, recent progress in the development of organisms that accumulate bio-oils (i.e. triesters of fatty acids, FAs, and glycerol) and the abundance of glycerol- and fatty acid-rich industrial byproducts make these carbon sources an attractive alternative. In addition to abundance, the higher degree of reduction and higher metabolic efficient of glycerol and FAs can support product yields significantly higher than those obtained from sugars. However, most industrial organisms metabolize glycerol and FAs only under respiratory conditions, which do not lead to the synthesis of fermentation products. We then have used synthetic and systems biology approaches to engineer the efficient synthesis of fuels and chemicals from these substrates (references below). Several native and heterologous fermentative pathways were engineered to function in E. coli under anaerobic and microaerobic conditions. Pathways for the efficient utilization of glycerol and FAs were engineered as well. Biofuels (ethanol, butanol, and hydrogen) and biochemicals (succinate, lactate, acetate, propionate, 1,2-propanediol, acetone, and isopropanol) were chosen as representative products to illustrate the feasibility of the proposed approach. Product yields in most cases exceeded those achieved with the use of sugars, clearly demonstrating the advantages of glycerol and FAs.

- Dellomonaco et al. 2009. Synthetic respiro-fermentative metabolism for the production of fuels and chemicals from fatty acids. App Environ Microbiol (MS in Review). - Gonzalez et al. 2009. Production of ethanol from thin stillage by metabolically engineered Escherichia coli. Biotechnol Letters (DOI: 10.1007/s10529-009-0159-2). - Gupta et al. 2009. Anaerobic fermentation of glycerol in Paenibacillus macerans: metabolic pathways and environmental determinants. App Environ Microbiol 75: 5871. - Durnin, et al. 2009. Understanding and harnessing the microaerobic metabolism of glycerol in Escherichia coli. Biotechnol Bioeng 103: 148. - Yazdani and Gonzalez. 2008. Engineering Escherichia coli for the efficient conversion of glycerol to ethanol and co-products. Metab Eng 10: 340. - Gonzalez et al. 2008. A new model for the anaerobic fermentation of glycerol in enteric bacteria: trunk and auxiliary pathways in Escherichia coli. Metab Eng 10: 234. - Murarka et al. 2008. Fermentative utilization of glycerol in Escherichia coli and its implications for the production of fuels and chemicals. App Environ Microbiol 74: 1124. - Yazdani and Gonzalez. 2007. Anaerobic fermentation of glycerol: A path to economic viability for the biofuels industry. Curr Opin Biotechnol 18: 213. - Dharmadi et al. 2006. Anaerobic fermentation of glycerol by Escherichia coli: a new platform for metabolic engineering.

Biotechnol Bioeng 94: 821.

See more of this Session: Advances in Metabolic Engineering and Bioinformatics I - Biofuels
See more of this Group/Topical: Food, Pharmaceutical & Bioengineering Division

(380b) Engineering Ethanol Tolerance in *Escherichia coli* for Improved Ethanol Production

Lauren B. Andrews, Brian L. May, Joseph R. Warner and Ryan T. Gill, Chemical and Biological Engineering, University of Colorado at Boulder, Boulder, CO

Recombinant *Escherichia coli* has been shown to be a potential strain for commercial cellulosic ethanol production with its ability to efficiently convert both pentose and hexose sugars into ethanol. However, native *E. coli* lacks many desirable traits for bioethanol production, including high tolerance to ethanol. In this work, we have identified genes which confer improved tolerance to ethanol using plasmid-based genomic libraries and selections. The most highly enriched genes identified by microarray were cloned and tested for improved growth with exogenous ethanol (growth rate and minimum inhibitory concentrations). In order to investigate the relationship between improved growth with exogenous ethanol and improved ethanol production, we tested these clones for improved ethanol production with inducible heterologous expression of the *Zymomonas mobilis* homoethanol production pathway. This *E. coli* ethanol production strain was constructed with a strong, tightly regulated promoter and the ethanol production cassette was inserted at a genomic region with minimal polar effects. This work identifies promising genetic targets for further ethanol tolerance engineering and provides insight into using genome-wide approaches for strain engineering improved production of inhibitory biofuels.

See more of this Session: Advances in Metabolic Engineering and Bioinformatics I - Biofuels
See more of this Group/Topical: Food, Pharmaceutical & Bioengineering Division

(380c) Engineering A Synthetic Microbial Consortium for Efficient Production of Biofuels

Jeremy J. Minty, Marc Singer, Alissa Kerner, Jungho Ahn and Xiaoxia Lin, Department of Chemical Engineering, University of Michigan, Ann Arbor, MI

Production of liquid biofuels via microbial conversion of cellulosic feedstocks could sustainably replace much of the fossil fuel consumed in the transportation sector, but tremendous research efforts are still needed in engineering microorganisms with the desired properties. Inspired by the widespread occurrence of synergistic microbial communities in nature, we are investigating a promising but under-explored direction for engineering microorganisms for biofuel production: the design and construction of a synthetic microbial consortium, composed of various microorganisms, each specialized for a specific task in the conversion of lignocellulosic biomass to a biofuel. Three microbial specialists are utilized for the consortium: the cellulolytic fungus *Trichoderma reesei* RUT-C30, which secretes cellulase enzymes to hydrolyze lignocellulose into component hexose and pentose mono/oligosaccharides, an engineered *E. coli* hexose specialist that ferments hexose hydrolysis products, and an engineered *E. coli* pentose specialist that ferments pentose products. In contrast to the "superbug" paradigm of trying to engineer one organism with all the necessary functionalities to produce biofuels from cellulose, our approach can potentially be more tractable from an engineering and feasibility standpoint. In our initial work, we focus on developing the cellulose utilization aspect of the consortium. The modular nature of our system will allow it to be readily adapted to the large portfolio of existing *E. coli* strains metabolically engineered to produce biofuels or commodity chemicals.

Lignocellulose is highly recalcitrant to biological degradation [1]. Biofuel production hosts such as *S. cerevisiae* have been engineered to ferment cellulose via heterologous cellulase production, but reported cellulose hydrolysis rates and yields are modest [2]. In contrast, natural saprophytic organisms have evolved highly specialized and sophisticated synergistic enzyme systems for efficient degradation of cellulose [1]. The filamentous fungus *Trichoderma reesei* RUT-C30 is a well characterized, prodigious cellulase producer [1] and is utilized as a base strain for the cellulolytic specialist in our consortium. Antagonistic interactions between filamentous fungi and bacteria are commonly observed in nature, but we have successfully co-cultured *E. coli K12* and *T. reesei* RUT-C30 on minimal medium with crystalline cellulose (Avicel) as the sole carbon source, demonstrating the feasibility of the consortium concept.

Diauxic sugar utilization in *E. coli* reduces fermentation efficiency, motivating the development of hexose and pentose specialist E. coli strains [3]. A pentose specialized *E. coli* was developed by disabling hexose metabolism via a series a gene deletions (*ptsG, glk,* and *manX*); presently we are working to improve utilization of pentose sugars, which are metabolized less efficiently than hexose sugars. For the hexose specialist, our efforts are focused on engineering this strain to utilize cellooligosaccharides, which are primary products of enzymatic cellulose hydrolysis and are not natively metabolized by *E. coli*. The cryptic *chbBCARF* operon of *E. coli* encodes a functional cellobiose permease and phospho-β-glucosidase, but is constitutively repressed in most laboratory strains [4]. We have engineered a cellobiose utilizing strain by expressing the *chbBCARF* operon on a plasmid with a heterologous promoter and are working to further improve cellobiose metabolism.

Preliminary modeling work indicates that the populations of each consortium member must be balanced for maximum biofuel production, and that the optimal consortium composition may vary with the cellulosic feedstock concentrations and makeup. Although the sub-populations can be controlled by varying the inoculation ratio of the consortium members, the system would be much more robust if the organisms could be coordinated and regulated with external signals (e.g. inducers) or even those internal in the consolidated bioreactor (e.g. feedstock composition). To address this need, we are also engineering synthetic metabolic and genetic circuits to allow organisms to sense and respond to population levels and available substrates.

[1] Lynd, Lee R., Willem H. Van Zyl, John E. McBride, and Mark Laser. "Consolidated Bioprocessing of Cellulosic Biomass: an Update." Current Opinion in Biotechnology 16.5 (2005): 577-83.

[2] Wen, Fei, Jie Sun, and Huimin Zhao. "Yeast Surface Display of Trifunctional Minicellulosomes for Simultaneous Saccharification and Fermentation of Cellulose to Ethanol." Applied and Environmental Microbiology 76.4 (2010): 1251-260.

[3] Eiteman, Mark A., Sarah A. Lee, and Elliot Altman. "A Co-fermentation Strategy to Consume Sugar Mixtures Effectively." Journal of Biological Engineering 2.3 (2008).

[4] Kachroo, Aashiq H., Aswani K. Kancherla, Nongmaithem S. Singh, Umesh Varshney, and Subramony Mahadevan. "Mutations That Alter the Regulation of the Chb Operon of Escherichia Coli Allow Utilization of Cellobiose." Molecular Microbiology 66.6 (2007): 1382-395.

See more of this Session: Advances in Metabolic Engineering and Bioinformatics I - Biofuels
See more of this Group/Topical: Food, Pharmaceutical & Bioengineering Division

(380d) Metabolic Engineering of Clostridium Tyrobutyricum for Butanol Production

Mingrui Yu, Department of Chemical Engineering, The Ohio State University, Columbus, OH, Yali Zhang, The Ohio State Biochemistry Program, The Ohio State University, Columbus, OH and Shang-Tian Yang, Department of Chemical and Biomolecular Engineering, The Ohio State University, Columbus, OH

Compared to ethanol, butanol has gained more and more interest as a substitute of gasoline because of its higher energy content, lower volatility and less corrosion. Butanol can be produced by anaerobic microorganisms such as Clostridium acetobutylicum and C. beijerinckii in acetone-butanol-ethanol fermentation (ABE fermentation), which however suffers from a low butanol yield (~20% w/w), low titer (<15 g/L) and low productivity because of butanol toxicity and the production of other byproducts including acetone, ethanol, acetate and butyrate. Conventional ABE fermentation involves complicated metabolic pathways, including acidogenesis, solventogenesis and spore-forming life cycle, which are difficult to control or manipulate. Despite extensive efforts in metabolic engineering of solventogenic Clostridia, little progress has been made in improving butanol yield, titer, and productivity in ABE fermentation. In this work, we aim at developing a novel solventogenic Clostridia strain by cloning and expressing genes in the butanol biosynthesis pathway into C. tyrobutyricum, a butyric acid bacterium that normally produces butyric acid with acetic acid, carbon dioxide and hydrogen as byproducts from glucose and xylose. C. tyrobutyricum has high tolerance to butyric acid (>50 g/L) and butanol (>20 g/L). It's relatively simple metabolic pathway and high metabolic flux to butyryl-CoA (the precursor for butanol biosynthesis) render it an ideal host for butanol production.

In this study, several butanol producing mutants of C. tyrobutyricum were obtained by overexpressing aldehyde-alcohol dehydrogenase (adhE2) under the control of native thiolase promoter. A butanol titer of 1.1 g/L was obtained in batch fermentations carried out in serum bottles with wild-type C. tyrobutyricum overexpressing adhE2 (WT-ad). Higher butanol titers were achieved in fermentations with ptb (phosphotransbutyrylase) and ack (acetate kinase) knockout mutants. The ack knockout mutant (ACK-ad) produced the highest butanol (~10 g/L) with a high yield (0.26-0.33 g/g glucose consumed). Butyric acid was the major byproduct along with some acetic acid and ethanol in the fermentation. Compared to WT-ad, the ack knockout mutant (ACK-ad) produced more C4 products (butanol and butyrate) and less C2 products (acetate and ethanol) with a high C4/C2 ratio of 10.6 (mol/mol), whereas the ptb knockout mutant (PTB-ad) produced more C2 and less C4 products with a low C4/C2 ratio of 1.4. It is clear that ack and ptb genes play important roles in controlling the metabolic flux distribution in C. tyrobutyricum. These results suggest that knockout of genes in the butyrate biosynthesis pathway in the mutant ACK-ad can further increase the butanol yield to approaching the theoretical value of 0.41 g/g glucose.

See more of this Session: Advances in Metabolic Engineering and Bioinformatics I - Biofuels

(380e) Improving Exogenous Sugar Utilization: a Survey of Xylose Molecular Transporters

Eric M. Young, Chemical Engineering, University of Texas at Austin, Austin, TX and Hal Alper, Chemical Engineering, The University of Texas at Austin, Austin, TX

Metabolic engineering is a powerful platform for improving cellular production of biomolecules. However, classical metabolic engineering tools do not address molecular transport across the cellular membrane – the key first step in most pathways. Molecular transport exerts limited flux control when an organism utilizes its preferred carbon source, yet its limitation is clearly evident when altering the carbon source preferences of an organism. As an example, molecular transport limits pentose utilization in *Saccharomyces cerevisiae* , an important problem in lignocellulosic biomass conversion to biofuels. Due to the lack of a native xylose transport protein, *S. cerevisiae* transports xylose via its hexose transporters where low affinity and hexose inhibition cause low flux. To improve this utilization, xylose transporters in yeast must be imported and/or engineered. Here, we present the first step towards this goal, namely a survey of a large group of known and putative xylose transporters from yeast, bacteria, and plant. These transporters were cloned and expressed in recombinant xylose-utilizing *S. cerevisiae* . The conferred growth on xylose and glucose was measured, along with an analysis of substrate preference. This study provides insight into the relative performance of these heterologous proteins as well as candidates for further investigation and improvement. Moreover, this work establishes the basis for molecular transporter protein engineering.

See more of this Session: Advances in Metabolic Engineering and Bioinformatics I - Biofuels
See more of this Group/Topical: Food, Pharmaceutical & Bioengineering Division

(380f) Molecular Engineering of Xylose Transport and Metabolism in Escherichia Coli and Saccharomyces Cerevisiae

Tingjian Chen, Chuan Ren, Ling Liang and Zhanglin Lin, Department of Chemical Engineering, Tsinghua University, Beijing, China

Efficient utilization of xylose is necessary for economic production of biochemicals and biofuels from lignocellulosic materials[1]. Escherichia coli and Saccharomyces cerevisiae are two of the most important workhorses for bio-refinery of biomass. Xylose transport and metabolism process are important for their efficient use of xylose in the lignocellulosic hydrolysate[2-4]. In the presence of glucose, the xylose transport and metabolism in microorganisms is often inhibited, due to the carbon catabolite repression (CCR) mechanism. In this work, we introduced a heterologous xylose transporter, glf from Zymomonas mobilis, into E .coli, and carried out directed evolution of this transporter, using a high throughput assay of xylose transport[5]. After 2 rounds of evolution, we obtained a mutant RD5, the xylose transport activity of which increased more than 7 folds in the presence of glucose. Fermentation tests showed that RD5 significantly improved sugar utilization in media supplemented with xylose or a mixture of xylose and glucose, compared with the wild type strain. To make S. cerevisiae be able to metabolize xylose efficiently, XylA, xylose isomerase from E. coli was cloned into S.cerevisiae, and then evolved by error-prone PCR and selection on plates containing xylose as a sole carbon source. After two rounds of selection, a mutant 1-M13 was obtained. Fermentation tests showed that the xylose utilization for this mutant improved to about 1.7 folds of the wild type strain in media supplemented with mixtures of xylose and glucose. The xylose isomerase activity profile showed that the optimal pH value for 1-M13 has shifted to a more acidic range than the wild type.

References: 1. Stephanopoulos, G., Challenges in engineering microbes for biofuels production, Science 2007. 315: 801-804. 2. Jeffries, T.W., Engineering yeasts for xylose metabolism. Curr Opin Biotechnol, 2006. 17(3): 320-6. 3. Henderson, P.J., Proton-linked sugar transport systems in bacteria. J Bioenerg Biomembr, 1990. 22(4): 525-69. 4. Saloheimo, A., et al., Xylose transport studies with xylose-utilizing Saccharomyces cerevisiae strains expressing heterologous and homologous permeases. Appl Microbiol Biotechnol, 2007. 74(5): 1041-52. 5. Chen, T., et al., An in-vivo, label-free quick assay for xylose transport in Escherichia coli, Anal Biochem, 2009, 390 (1): 63-67.

See more of this Session: Advances in Metabolic Engineering and Bioinformatics I - Biofuels
See more of this Group/Topical: Food, Pharmaceutical & Bioengineering Division

(380g) Improving Ethanol Production of Xylose-Utilizing Saccharomyces Cerevisiae

Ljubisa Miskovic, Laboratory of Computational Systems Biotechnology, Ecole Polytechnique Fédérale de Lausanne (EPFL), Lausanne, Switzerland and **Vassily Hatzimanikatis**, Institute of Chemical Sciences and Engineering, Ecole Polytechnique Fédérale de Lausanne (EPFL), Lausanne, Switzerland

Abundant in nature lignocellulosic plant biomass represent a promising, sustainable, source of energy as it can be converted into fuel ethanol through the fermentation of one of its prevalent sugar constituents – xylose. Xylose-utilizing recombinant Saccharomyces cerevisiae strains are engineered to overcome the potential bottlenecks of xylose uptake and cofactor imbalance, and to minimize the xylitol accumulation. In this work, we apply a novel computational framework to identify metabolic engineering targets for the optimization of ethanol production by Saccharomyces cerevisiae in the fed-batch culture. We employ a methodology that uses Monte Carlo sampling to explore the kinetic data space, and it accounts explicitly for mechanistic properties of enzymes and physico-chemical and thermodynamic constraints. We model the kinetic responses of the individual reactions, and we integrate this mechanistic information with the thermodynamically feasible pairs of substrate and product concentrations to build a population of all biochemically and thermodynamically meaningful models of the Xylose-utilizing recombinant Saccharomyces cerevisiae strains. We find that redistribution of the metabolic flows around pyruvate among carboxylation, decarboxylation, and transport has significant potential in improving total utilization of hexose and pentose sugar as well as its conversion into ethanol.

See more of this Session: Advances in Metabolic Engineering and Bioinformatics I - Biofuels
See more of this Group/Topical: Food, Pharmaceutical & Bioengineering Division

(399a) From Metabolic Modelling to Integrated Bio-Refineries: How to Improve the Sustainability of Biodiesel Plants by Co-Production of Chemicals

Anestis Vlysidis[1], Michael Binns[1], Colin Webb[2] and Constantinos Theodoropoulos[3], (1)Chemical Engineering and Analytical Science, University of Manchester, Manchester, United Kingdom, (2)School of Chemical Engineering and Analytical Science, The University of Manchester, Manchester, United Kingdom, (3)School of Chemical Engineering and Analytical Science, University of Manchester, Manchester, United Kingdom

Biofuels, such as bioethanol and biodiesel, which are alternatives to conventional fuels, have recently received increasing interest. However, costly raw materials and inefficient utilization techniques of the waste streams of these industries have resulted in being less competitive to petroleum refineries. The aim of this project is to utilize the side products (mainly glycerol) of biodiesel's industry in an integrated and cost effective point of view. Biodiesel is a transport fuel made from vegetable oils or animal fats through a chemical reaction process called transesterification that co-produces glycerol as by-product (10% w/w) which cannot be absorbed by the present glycerol markets [1].

In this study, we introduce a new and efficient bio-process of succinic acid production from glycerol. Initially, an unstructured model is developed for the prediction of relevant batch experiments that we have performed, which considers both substrate and product inhibition. Based on model' kinetics we examine the best conditions on batch experiments in terms of productivity. In addition to that, we perform an optimal control analysis investigating the best feeding profiles for fed-batch processes in order to maximize the productivity of the desired product. In both cases, computational results are validated with our batch fermentation experiments. Moreover, we try to give further insights to this bacterial process through metabolic modelling. We study Actinobacillus succinogenes' metabolic network and underline the dominate pathways and the main branch points that affect the succinate production. We implement flux balance analysis to compute the unknown intracellular fluxes and linear optimisation techniques to maximise further the productivity. Finally an integrated bio-refinery framework involving a cradle to grave analysis is combined with the above experimental and computational findings. Overall simulations and life cycle analysis of the entire process illustrate the "beneficial" uses of glycerol in terms of economics and sustainability.

References

1. Johnson DT, Taconi KA, Environmental Progress 26:338-348, (2007)

(399b) A Decision Support Platform for the Configuration of Biomass Based Production Systems

Marinella Tsakalova[1], Ta-Chen Lin[1], Aidong Yang[1] and **Antonis Kokossis**[2], (1)Faculty of Engineering and Physical Sciences, University of Surrey, Surrey, United Kingdom, (2)School of Chemical Engineering, National Technical University of Athens, Athens, Greece

The depletion of fossil resources and the ever increasing demand on the reduction of GHG emissions have driven the process industry towards incorporating renewable feedstock, i.e. biomass, for producing energy, chemicals, and materials. Given the vast amount of possibilities in selecting the types of biomass feedstock, conversion technologies, and target products, a systematic approach is highly desirable to support the decision making process. Along this direction, tools including simulation analysis and process synthesis and integration have already been applied to the investigation and optimisation of individual plants or processes for biomass conversion.

As opposed to the study of individual processes, the Department of Environment, Food and Rural Area (DEFRA) in UK has supported the development of a systematic approach to screen biomass feedstocks and design options that may involve multiple interconnected processes. In this work, a Unit of Synthesis (UoS), typically representing a complete conversion process, e.g. that of bioethanol or biodiesel production, can be connected with other UoS to form a production chain, which is referred to as a processing route. An example of a two-step processing route is the production of ethanol by fermentation (UoS 1) followed by a dehydration process (UoS 2) to eventually produce ethylene. Furthermore, multiple feasible routes, often involving common processing steps or UoS, may be formulated into a superstructure to represent all the possibilities for converting a set of feedstocks to a set of products. The optimisation of the superstructure against economical, energy, and/or GHG objectives will allow to systematically identify promising configuration(s) of the production system, represented by the best combination(s) of chosen feedstocks, processing steps, and products.

The above functionalities are currently being implemented in a systems platform to meet different levels of industrial needs. The platform consists of a central spreadsheet-based database, a spreadsheet-based simulation tool to allow the evaluation of pre-defined processing routes, and a GAMS-based synthesis tool to perform superstrcuture-based optimal screening/configuration. The contral database holds data of individual Units of Synthesis and other common data such as prices of chemicals/utilities. This database is shared by both the simulation tool and the synthesis tool.

The DEFRA-funded project aims to study eight chemicals, which include two key chemicals namely ethylene and propylene glycol. A number of routes have been identified for producing each of the two key chemicals. The systems platform will be tested in the first pace by case studies addressing these two chemicals separately or jointly.

See more of this Session: Environmental Biotechnology: Green Bioprocessing
See more of this Group/Topical: Food, Pharmaceutical & Bioengineering Division

(399c) Novel Self-Assembled Protein Nanostructures for Xylan Bioprocessing

Shara McClendon, Hyun-Dong Shin, Zichao Mao and Rachel Ruizhen Chen, Chemical & Biomolecular Engineering, Georgia Institute of Technology, Atlanta, GA

Lignocellulose, present in numerous industrial waste and agriculture residues, can be used for production of a wide variety of molecules including biofuels, commodity chemicals and materials. However, processing the recalcitrant material in a cost-effective and environmental-friendly manner presents significant engineering challenges. Bioprocessing of hemicellulose is one example that illustrates such challenges. Its complete hydrolysis requires not only backbone-acting enzymes but also the synergistic action of several side-chain acting enzymes. To effectively hydrolyze the target substrates, cellulolytic aerobic microbes secrete copious amounts of hemicellulases into the extracellular environment, whereas anaerobic organisms organize enzymes in cell-associated, self-assembled protein nanostructures, known as cellulosomes. Cellulosomes are constructed using a scaffoldin protein with repeating cohesin domains and enzymes tagged with corresponding dockerins. The specific protein-protein interaction between cohesins and dockerins provide the method of self-assembly. These protein structures position multiple proteins in close proximity to the substrate, resulting in greater enzymatic synergy and improved catalytic efficiency compared to unorganized, free enzymes.

Mimicking the architecture of cellulosomes, we constructed four two-unit nanostructures named xylanosomes for enhancing xylan hydrolysis. Four different enzymes with either a *Clostridium thermocellum* or *Clostridium cellulovorans* dockerin domain were cloned and expressed in *E. coli* to cover four out of the six required activities to fully hydrolyze xylan: xylanase, feruloyl esterase, arabinofuranosidase, and beta-xylosidase. These enzymes were combined with a complementary chimeric two-cohesin scaffoldin protein to create four possible enzyme combinations. The performance of both free enzyme mixtures and self-assembled structures was investigated on a hemicellulose substrate, wheat arabinoxylan. Results reveal that select enzyme combinations show synergy in these structures giving up to 3 times greater release of reducing sugars compared to a free system. Conversely, some structures gave negative synergies or showed no improvement. Additionally, different synergies were observed in combining either family 10 or family 11 xylanases with feruloyl esterases.

The presentation will detail the construction of these xylanosomes and their ability to selectively improve xylan hydroysis compared to the corresponding free enzyme system. Possible explanations for the observed synergies or lack thereof will be discussed.

See more of this Session: Environmental Biotechnology: Green Bioprocessing
See more of this Group/Topical: Food, Pharmaceutical & Bioengineering Division

(399d) Lipid Accumulation Kinetics and Microbial Community Analysis of Activated Sludge Microbiota

Andro Mondala[1], Rafael Hernandez[1], W. Todd French[1], Jorge Santo Domingo[2], Hodon Ryu[2], Linda McFarland[1] and Brandon Iker[2], (1)Chemical Engineering, Mississippi State University, Mississippi State, MS, (2)US Environmental Protection Agency, Cincinnati, OH

The prospect of utilizing the microbiota in sewage activated sludge for the production of lipids for biofuels application via fermentation of lignocellulose sugars was investigated as an alternative to plant-based lipid feedstocks. In addition to this, there are no known studies dealing with the relationship between lipid production and microbial composition of sludge systems in which lipid yields are enhanced by adjusting the carbon-nitrogen (C:N) ratio of the wastewater. In this study, the effect of changing the C:N ratio (10:1, 40:1, 70:1) and initial sugar loading (20, 40, 60 g/L) on activated sludge bioreactors dosed with glucose was examined for seven days. Biomass production, lipid accumulation (as fatty acid methyl esters), and glucose uptake kinetics were measured using conventional gravimetric and instrumental methods. Variations in the microbial composition of the activated sludge were determined using sequence analysis of 16S rRNA gene clone libraries. Results show that lipid production was higher in high C:N ratio bioreactors, although the total cell biomass and sugar utilization were lower. After 48 h, *Firmicutes* (Gram positive bacteria) represented two thirds of the bacteria in moderate and low C:N ratio reactors while *Proteobacteria* (Gram negative bacteria) represented 87 % of the bacteria in reactors exposed to high a C:N ratio. *Proteobacteria* membership also changed between 48 h, with gamma-*Proteobacteria* representing over 82 % of all bacteria, and 168 h with alpha-*Proteobacteria* representing over 86 % of all bacteria in high C:N ratio bioreactors. The results show that changes in microbial composition may be correlated to lipid production. These data will be used in future studies to identify and isolate potential lipid-accumulating bacteria in municipal wastewater activated sludge.

See more of this Session: Environmental Biotechnology: Green Bioprocessing
See more of this Group/Topical: Food, Pharmaceutical & Bioengineering Division

(399f) Single Cell Protein Production From Glycerol Residues Discharged From Biodiesel Production Plant

Nuttha Thongchul[1], Sajee Noitang[1], Sarintip Sooksai[1], Amorn Petsom[1], Veerapat Tantayakom[2] and Phatthanon Prasitchoke[2], (1)Institute of Biotechnology and Genetic Engineering, Chulalongkorn University, Bangkok, Thailand, (2)PTT Chemical Public Company Limited, Bangkok, Thailand

Glycerol is now considered as the major byproduct from biodiesel manufacture. Nowadays, the application of glycerol is very limited. In the near future, biodiesel production may suffer from glycerol surplus if there is no sustainable solution for glycerol utilization. In a large commercial biodiesel plant, crude glycerol often enters purification process which eventually yields highly purified glycerol for applications in pharmaceutical and cosmetic industries. Nonetheless, a large amount of purified glycerol is still a surplus for pharmaceutical and cosmetic industries. Using the biotechnology technique to convert glycerol to the value-added product has been brought into attraction. It is well known that *Hansenula polymorpha* consume glycerol for both carbon and energy sources. Under aerobic condition, glycerol is usually converted by glycerol kinase into glycerol-3-phosphate (G3P), which is then oxidized to dihydroxyacetone phosphate (DHAP) by a membrane-bound G3P dehydrogenase before entering a central metabolic pathway. In this work, batch fermentation of *H. polymorpha* IBGE HP-5001 in YE-glycerol medium using different glycerol samples discharged from purification process of biodiesel production plant has been studied in a 5-L stirred tank bioreactor. It was found that *H. polymorpha* IBGE HP-5001 was capable of consuming glycerol in glycerol containing discharged streams for biomass production. The highest cell biomass production was observed in the fermentation of YE-glycerol media containing 5% pharma glycerol and 5% technical glycerol while low biomass yield was obtained in the fermentation of YE-glycerol media containing 50% residual glycerol and 9% crude glycerol due to high osmolarity. In order to reduce the production cost, the medium containing yeast extract as an enriched organic N-source was replaced by the minimal salts medium. Due to a large amount of salts remained untreated in the medium, an inorganic N-source ($(NH_4)_2SO_4$) with supplemented salts ($MgSO_4$ and KH_2PO_4) even caused higher osmotic pressure as compared to those in the YE-glycerol medium. This resulted in low biomass production. Salt removal from discharged glycerol is recommended to obtain the higher cell biomass production.

See more of this Session: Environmental Biotechnology: Green Bioprocessing
See more of this Group/Topical: Food, Pharmaceutical & Bioengineering Division

(399g) Transcriptional Response of Lactobacillus Brevis to Sudden Ferulic Acid Stress

James D. Winkler, Chemical Engineering, Texas A&M University, College Station, TX and Katy Kao, Texas A&M University, College Station, TX

There is great interest in the use of waste biomass to reduce the environmental footprint and cost of biological processes. Preparing the biomass by acid hydrolysis generates toxic phenolic acids such as the model compound ferulic acid, however, that must be removed prior to further downstream processing. Lactobacillus brevis, a poorly characterized heterofermentative lactic acid bacterium, exhibits high tolerances of ferulic acid and other phenolic acids that are common in hydrolyzed biomass. A better understanding the genetic basis for this complex phenotype can guide efforts to engineer other bacterial strains. Here we examine the transcriptional response of L. brevis to sudden ferulic acid stress to achieve that end. We find evidence to suggest that L. brevis undertakes a variety of ameloriative actions, including specific membrane modifications, phenolic acid decarboxylase expression, and others in response to ferulic acid in situ. These results will aid in the creation of phenolic-tolerant strains necessary for the utilization of untreated biomass hydrolysates.

See more of this Session: Environmental Biotechnology: Green Bioprocessing
See more of this Group/Topical: Food, Pharmaceutical & Bioengineering Division

(678a) Catalysis in Flow: Practical and Selective Transformations of Alcohols

Mimi Hii[1], Natalia Zotova[1], Klaus Hellgardt[2] and Geoff Kelsall[2], (1)Chemistry, Imperial College London, London, United Kingdom, (2)Chemical Engineering, Imperial College London, London, United Kingdom

Development of clean, selective, atom-efficient reactions has become the new paradigm in organic synthesis, driven by a desire to reduce economic, as well as environmental costs of chemical production. In this respect, one of the most challenging reactions is the oxidation of alcohols to aldehydes and ketones, which is ubiquitous in organic chemistry, yet also one of the least efficient. The lack of chemoselective and atom-economical oxidants often make these reactions unattractive, particularly for the large scale synthesis of complex molecules. In this presentation, we will demonstrate a new way of transforming alcohols to other functional molecules by performing catalysis in a continuous flow reactor which fulfil several criteria for practical and 'green' synthesis

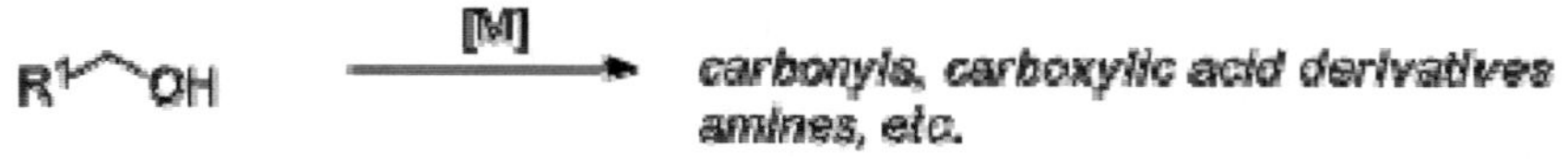

See more of this Session: Application of Green Chemical Engineering Approaches to Improve Sustainability in Pharmaceutical Manufacturing
See more of this Group/Topical: Food, Pharmaceutical & Bioengineering Division

(678b) Combining Homogeneous Reactions with Heterogeneous Separations for the Hydroformylation of Styrene Using Tunable Solvents

Ali Z. Fadhel[1], Jackson W. Ford[2], Andy Wu[1], Veronica Llopis Mestre[1], Pamela Pollet[3], Charles L. Liotta[3] and Charles A. Eckert[1], (1)School of Chemical & Biomolecular Engineering, Georgia Institute of Technology, Atlanta, GA, (2)Center for Environmentally Beneficial Catalysis and Department of Chemical & Petroleum Engineering, University of Kansas, Lawrence, KS, (3)School of Chemistry & Biochemistry, Georgia Institute of Technology, Atlanta, GA

Organic Aqueous Tunable Solvents combine the benefits of homogeneous and heterogeneous catalysis while overcoming the limitations of both. We are using the hydroformylation of hydrophobic p-methylstyrene as a model chiral reaction, which mimics the active pharmaceutical ingredient in ibuprofen production. The homogeneous hydroformylation reaction is carried out in organic-water mixtures using Rh/TPPMS catalytic complex under 3 MPa of syngas. Reaction rates are improved by an order of magnitude and the yield is increased by 20% when compared to reported heterogeneously catalyzed reactions. After the reaction, moderate pressures of carbon dioxide (around 3 MPa) are used to induce a phase split, as the gas is almost insoluble in water but completely miscible with most organic solvents. The hydrophobic product partitions into the organic-rich phase, which is decanted and then depressurized to yield a facile separation of >99% recovery. The aqueous-rich phase, which contains 99.9% of the hydrophilic catalyst, is recycled for the next reaction cycle. We show five recycles of the catalyst without significant loss of activity.

See more of this Session: Application of Green Chemical Engineering Approaches to Improve Sustainability in Pharmaceutical Manufacturing
See more of this Group/Topical: Food, Pharmaceutical & Bioengineering Division

(678c) Green Engineering Analysis of a Multi-Use Solvent Recovery System for Small Volume Waste Steams in the Pharmaceutical Industry

C.Stewart Slater[1], Mariano J. Savelski[1], Frank J. Urbanski[2], Gregory Hounsell[2], Joseph W. Geiger[3] and Donald J. Knoechel[3], (1)Chemical Engineering, Rowan Univeristy, Glassboro, NJ, (2)Pfizer Global Manufacturing, Pfizer, Peapack, NJ, (3)Pfizer Global Manufacturing, Pfizer, Kalamazoo, MI

A green engineering study has been conducted to analyze the economic viability and quantify the environmental benefits of a small solvent recovery system as an alternative to incineration for addressing smaller volume waste streams.

Solvent recovery is a routine practice in the pharmaceutical industry when it is technically and economically viable for the particularly waste stream. Capital investment in the required piping, tank farms and recovery equipment is more easily justified when dealing with large volumes, high cost solvents and high equipment utilization rates, and when solvents from multiple products can be pooled together – i.e. the don't require segregation by product. The use of recovered solvents, and pooling of solvents, must be appropriately qualified to assure product quality and avoid cross contamination. Economic justification to recover small-volume, "non-poolable", and intermittently generated waste streams remains challenging, but a potential recovery opportunity.

A case study was performed on the waste streams being generated at an API synthesis building at the Pfizer Kalamazoo, Michigan plant. The goal was to investigate those streams that could be most easily recovered with simple phase or equilibrium-based separations. As a first step in that analysis, the recovery of acetonitrile from a waste stream in the selamectin synthesis was considered. Selamectin is the active ingredient in the veterinary drug Revolution®. This stream was initially chosen due to the relative high cost (and value) of acetonitrile and the ability to separate acetonitrile from the acetone impurity.

A small-scale distillation solvent recovery system was designed and the proposed operation compared with the current waste disposal practice. The case study estimates the environmental benefits, using life cycle assessment tools, and the financial metrics for the proposed improvement. To increase the economic feasibility of a capital investment and improve the environmental footprint further, the proposed design was evaluated for use with the other waste streams in the facility. The simulation included isopropanol recovery from the manufacture of nelfinavir, an antiretroviral drug (Viracept®); and toluene recovery from hydrocortisone manufacture.

An overall environmental analysis was performed evaluating the impact of utilities (and resulting emissions) needed to operate the distillation unit versus the savings from not having to manufacture the virgin solvent and incinerate it. These were then analyzed through a cost/benefit analysis.

See more of this Session: Application of Green Chemical Engineering Approaches to Improve Sustainability in Pharmaceutical Manufacturing
See more of this Group/Topical: Food, Pharmaceutical & Bioengineering Division

(678d) Pervaporation: An Alternative Process for the Dehydration of Pharmaceutical Streams

Dimitrios Zarkadas, Plamen Grigorov and Azzeddine Lekhal, Chemical Process Development & Commercialization, Merck and Co. Inc., Union, NJ

Pervaporation is a process that has been used successfully the last two decades for the dehydration of organic solvents at large scales and for improving distillation efficiency of azeotropic systems like ethanol-water. Similarly to other membrane processes, pervaporation offers several advantages. It often results in lower costs, it is modular and therefore easy to scale up, offers much higher mass transfer rates per unit volume compared to conventional separation equipment and can effect separations difficult to achieve otherwise. Despite the growing importance of pervaporation in the chemical industry, there are no reported efforts to apply the technology in the dehydration of pharmaceutical streams. The latter is important in several cases: removal of moisture during condensation reactions, generation of supersaturation during crystallization of intermediates or Active Pharmaceutical Ingredients (APIs) and recovery of solvent at a low energy cost. In this paper we report for the first time our efforts to dehydrate a stream containing a pharmaceutical intermediate. This intermediate is crystallized from a 2-MeTHF solution saturated with water. Moisture must be removed to levels lower than 0.3% wt. to avoid excessive yield losses during the crystallization process. The dehydrated solution is then distilled, cooled to below its saturation temperature, seeded and isolated. Prior to this study water removal was achieved by a distillation process. The latter required the consumption of significant amounts of solvent (6X) as well as prolonged processing time (about 30 hrs) at commercial scale. We have performed process development in a two-tier approach. We first evaluated the feasibility of the process at the laboratory scale by using ceramic modified silica membranes. Our laboratory runs clearly illustrated the feasibility of the process; water levels were reduced routinely to 0.15-0.25% wt. after only 2-3 hours of operation. Contrary to the previously used distillation process, no additional solvent was required to achieve the separation. We also assessed the long term performance of the membrane by operating the membrane system over an extended period of time. Continuous monitoring of moisture levels in the stream by an in-line KF instrument showed that the test was successful. Second, we successfully scaled up the process at our pilot plant. The KF specification was met and no significant operation issues were observed. The process is currently evaluated for implementation at the commercial scale. Our calculations indicate that significant savings can be realized. First, the savings by the reduced solvent usage can pay for the capital investment of purchasing the membrane modules after only 4-5 batches are performed. Second, operation time can be reduced to 18-24 hrs resulting in additional savings in operational costs and cycle time and throughput. Third, the process requires a much lower energy input, since the process stream is never heated to its reflux temperature as in the conventional distillation process. In conclusion, the developed pervaporation process can deliver the required separation goal in a robust manner, at a lower cost and reduce solvent usage resulting in a greener process.

See more of this Session: Application of Green Chemical Engineering Approaches to Improve

(678e) Solvent Recovery and Reuse in Manufacturing of Pharmaceutical API

Yubo Yang, Vikram Kalthod, Lei Cao and Domenico Marini, Abbott Laboratories, North Chicago, IL

It is desired to recover solvents from organic wastes of API manufacturing to make the process greener and more environmentally friendly. A significant cost-saving can be realized by recovering solvents from organic wastes and reusing them in API processes when the recovered solvents meet the appropriate specifications. It is technically feasible and economical to recover solvents from API processes using fractional distillation.

The solvents used for the API process are ethyl acetate and heptanes. Two organic wastes are generated in the process: wet ethyl acetate containing 2-4% of water and the crystallization mother liquors consisting of ethyl acetate and heptanes. The wet ethyl acetate is purified to remove water to make the recovered dry ethyl acetate suitable for the water sensitive reactions in the synthesis. It is technically challenging and costly to fully separate ethyl acetate and heptanes by fractional distillation as ethyl acetate and heptanes form an azeotrope. However, it is feasible to separate the azeotrope of ethyl acetate-heptanes from heptanes with a single fractional distillation column. This led us to develop a process to recover the azeotrope and heptanes from the mother liquor, and to justify that an equivalent quality of API can be produced using the recovered solvents. Overall, >90% of the solvents from the mother liquor can be recovered and reused.

After full implementation, the API process is very green as solvents from organic wastes are fully recovered and reused. Over 2 million liters of solvents will be recovered and reused annually for the API manufacturing.

See more of this Session: Application of Green Chemical Engineering Approaches to Improve Sustainability in Pharmaceutical Manufacturing
See more of this Group/Topical: Food, Pharmaceutical & Bioengineering Division

(678f) Process Development and Design for Greener Pharmaceutical Processes

Krist V. Gernaey, Pär Tufvesson and John M. Woodley, Department of Chemical and Biochemical Engineering, Technical University of Denmark (DTU), Kgs. Lyngby, Denmark

Pharmaceuticals are particularly difficult molecules to manufacture and a number of challenges need to be addressed for effective process development and design. First the target molecule needs to be made within specification using a process developed in a very short timeline, brought about by the limited patent-life of the product and the need to recover high discovery costs. The second challenge arises from the need to integrate new technologies, and in particular biotechnology. The majority of pharmaceuticals manufactured today (and in the pipeline) are small molecules, although an increasing number have more than one chiral center and are complex multi-functional molecules. Unsurprisingly, therefore biotechnology has a role in the synthesis of such molecules via fermentation (to supply starting materials), whole-cell biocatalysis (to carry out complex conversions involving enzymes which cannot be isolated or multiple steps) and immobilized enzymes (to create chiral center(s) using reusable catalysts). In all these cases the biotechnological production methods need to be integrated with multi-step chemical synthesis and purification technology in order to obtain a greener production process. The third challenge is to design processes with integrated control such that PAT can be fully implemented and processes can be operated more efficiently. Processes will only take full advantage of PAT concepts such as 'real time release' if traditional batch-processing based production is replaced by continuous production. This will mean that processes can be controlled on the basis of suitable in-line/on-line measurements, such that real-time control can ensure that the product is within specification and can be marketed without the need for extensive laboratory testing. In this presentation tools to address each of these three challenges will be presented with specific examples.

See more of this Session: Application of Green Chemical Engineering Approaches to Improve Sustainability in Pharmaceutical Manufacturing
See more of this Group/Topical: Food, Pharmaceutical & Bioengineering Division

(678g) Evaluating and Minimizing Water Usage and Wastewater Generation in Integrated Bioprocesses Using Process Simulation and Scheduling Tools

Demetri Petrides, INTELLIGEN, INC., Scotch Plains, NJ and Charles Siletti, INTELLIGEN, INC., Mount Laurel, NJ

Water of various qualities is consumed in large quantities in the biopharmaceutical industries. It is used for preparing media and buffer solutions and for equipment cleaning. It is also used for generating steam of various qualities. The supply of purified water and the treatment of wastewater water is a common bottleneck during throughput increase projects. This presentation shows how process simulation and scheduling tools may be applied to evaluate the demand for water at various levels of purity and easily size water supply and wastewater treatment systems. The study also examines the effect of moving portions of the process to single-use systems (disposables) that do not require cleaning and sterilization.

See more of this Session: Application of Green Chemical Engineering Approaches to Improve Sustainability in Pharmaceutical Manufacturing
See more of this Group/Topical: Food, Pharmaceutical & Bioengineering Division

AUTHOR INDEX

AUTHOR INDEX

American Institute of Chemical Engineers
3 Park Avenue
New York, New York 10016-5991

CURRAN ASSOCIATES INC
proceedings.com